# Palm Oil

# VĀG ABO NDS

Radical pamphlets to fan the flames of discontent
at the intersection of research,
art and activism.

Series editor: Max Haiven

Also available

004

# Palm Oil

## The Grease of Empire

Max Haiven

PLUTO PRESS

First published 2022 by Pluto Press
New Wing, Somerset House, Strand, London WC2R 1LA

www.plutobooks.com

British Library Cataloguing in Publication Data
A catalogue record for this book is available from the
British Library

ISBN  978 0 7453 4582 6   Paperback
ISBN  978 0 7453 4586 4   PDF
ISBN  978 0 7453 4584 0   EPUB

This book is printed on paper suitable for recycling and
made from fully managed and sustained forest sources.
Logging, pulping and manufacturing processes are
expected to conform to the environmental standards of
the country of origin.

Typeset by Stanford DTP Services, Northampton,
England

Simultaneously printed in the United Kingdom and
United States of America

VAG
ABO
NDS

# Contents

**PALM OIL**
*The Grease of Empire*

Artist: Amanda Priebe

# Whose grease?

We find ourselves in a system of racial capitalism that appears as a vast, globe-spanning system of mystified human sacrifice, hidden in plain sight. The stories of palm oil I want to tell you will trace this system's contours and seek answers in its past. These are a story of how one largely invisible thing emerged from the nexus of capitalism, colonialism, and empire to define the cruelties of our world. Secreted within it is a story of our collective power to transform the world for the better. The story of palm oil is our story. This almost magical and ubiquitous substance is part of the way our bodies reproduce themselves and the way our material world is reproduced. It is a key element in the vortex of labor, commodities, meaning-making, and social relationships that make up the world in which we both live. Palm oil binds us, revealing the space in between, the syntax of the world.

Nearly every element of the process that now finds you reading these words could have been touched or facilitated by palm oil:[1] it could be an additive in the paper, a stabilizer in the ink, or part of the resin in the binding of the book; it is almost certainly either inside or essential to the manufacture of one of the hundreds of the components of the digital electronic device on which I am typing these words, and on which you might be reading them. It's probable that one of the transport vehicles that conveyed these artefacts to you burned hydrocarbons that included palm oil-derived agrofuels. And it must be taken as given that the body and brain that writes and that reads has been reproduced, in part, through the metabolism of palm oil. We have both used palm oil products to clean or care for our skin. We have ingested palm oil as a carrier of medicines. Though I suspect neither of us are intentionally investors in the palm oil industry, we are nonetheless economically entangled with it. The money that we receive for our labor is blood in the same ocean. Though it derives from a natural source, we created refined palm oil as it exists today, and it has, in turn, helped to created us.

In the story of palm oil, we can catch a glimpse of the world as it is made and unmade. To read a world of palm oil as if it were our story is to recognize what connects us and what divides us. My hope is that in paying attention to palm oil we might exercise some shared narrative muscle, so long atrophied in this world of competitive individualism, so that some "we" emerges that can better know itself and act in concert to change our fate. If we made this world from palm oil, what else could we have made? What else might we yet make?

In the past, my work has been dedicated to trying to grasp what we don't understand about capitalism. We understand that it is a global system that organizes the energies of humans and my wager is that, in telling a story of palm oil, we can recognize that we live in what Ruth Wilson Gilmore calls "the age of human sacrifice."[4] And we can see how it emerges from a longer history of racial capitalism that makes some people vastly more susceptible to disposability than others.

By palm oil I am speaking of the derivative of specific palm plants, mostly *Elaeis guineensis*, the African oil palm, but also sometimes cultivated from its central American relatives, *Elaeis oleifera* and the more distant *Attalea maripa*. Oil palms are among the world's most bountiful and useful plants. *E. guineensis*, from which we get most of the world's palm oil, is native to West Africa, where people have cultivated and treasured it for centuries. From its marvelous saffron-colored seedpods (which, when ready to harvest, can weigh over 10kg) African people have for millennia derived not only cooking oil but also lamp oil, cosmetics, medicines, artistic materials, sacraments, and dyes. From its sap comes palm wine and a wide variety of remedies. From its leaves come roof thatching and arrow and spear shafts.[5] Yesterday and today, fragrant, fleshy, palm oil has served ceremonial and spiritual purposes in Africa and its diaspora. For many, red, virgin palm oil is the taste of home, the taste of family, the taste of history.

In the course of my research I had the pleasure of hearing stories of palm oil and *E. guineensis* from many who hail from West Africa. They universally told me of the great admiration people have for this ever-giving plant, how central it and its gifts are to ancient traditions and also to the spirit of innovation and creativity that has seen munities of Africans who liberated themselves from enslavement found in the local variant of the oil palm a familiar friend that supplied many of the necessities of a fugitive life.[7]

But, of course, the form of palm oil that you and I know is something else entirely: industrially produced derivatives of the palm fruit can, like a god with many faces, appear as some 200 different ingredients in nutritional, industrial, and cleansing products around the world.[8] RBD (refined, bleached, and deodorized) oil has become a staple of the diets of billions of people, especially poor people, around the world. This globally traded, indifferent commodity emerges from intensive processing plants, predominantly located in Indonesia and Malaysia, but also in West Africa and Latin America, typically on clear-cut or razed lands that once sustained rainforest, though they may have taken many other forms since. Fertilizer, pesticide, and herbicide used in the intensive cultivation of this cash crop has, more often than not, found its way into local waterways.[9] At these factories, as well as at the nearby plantations on which lab-germinated *E. guineensis* grow in neat rows, nine meters apart, most of the workers are, in one way or another, displaced, sometimes for multiple generations. Perhaps it was civil war, perhaps imperialist-backed counter-insurgency campaigns, perhaps the ecological impacts of mining, perhaps land grabbing, perhaps it was government or international "development" incentives seeking to relocate workers to locales more convenient for corporations in need of cheap labor.[10] As a result, palm oil workers are now typically dependent on precarious employment for a means of buying the necessities of life. Even those who nominally own the land they work find themselves ensnared in systems of exploitations.

Today, you and I find palm oil, palm kernel oil or derivatives of these substances in an estimated 50% of the world's supermarket foods, predominantly in industrially produced, processed foods like packaged baked goods, edible spreads, ramen noodles, dairy products, and snack foods. But palm oil also enters us in trace amounts in a mind-boggling diversity of preservatives, emulsifiers, stabilizers, coagulants and additives.[11] Palm oil's unique chemical composition and extreme cheapness makes it a perfect base or additive to industrially produced foods to afford a long shelf life and facilitate transit through globe-spanning networks of trade.[12] It covers us, too: it is in the lion's share of cosmetics (though some higher-end brands occasionally boast of avoiding it). It is an important element in the production of plastics, dyes, inks, paints, and even paper products, including product packaging. It is in many of the lozenges, pills, suppositories, and other consumer and professional medical products that we use to transform our bodies. And it is also present in a multitude of industrial and manufacturing products and processes, notably in the surfactants that are an important part of machine lubricants, dying and tinting processes, detergents, and a dizzying array of other processes.[13] Globally, 72 metric tons of refined palm oil was consumed in 2020, roughly 20 pounds per human being.[14] Its intensive cultivation has transformed our planet:

over 27 million hectares of the earth's surface is under palm oil cultivation, an area greater than the size of New Zealand and approximately equivalent to all the agricultural land in France.[15] The clearing of forest and especially peatland for palm oil cultivation adds significant quantities of carbon to the earth's atmosphere—an estimated 6% of global annual emission—contributing to the dire, if unevenly distributed, risks of climate change.[16]

How did this happen to us? Our story will necessarily begin with the origins of the global commodity of palm oil in the European colonization of West Africa in the nineteenth century, where whole civilizations and millions of lives were sacrificed on the altar of that three-faced god: capitalist accumulation, white supremacist ideology, and inter-imperialist rivalry. We will visit nineteenth century Liverpool, where palm oil literally and figuratively lubricated the wheels of empire and furnished rich and poor alike with new commodities like soap. We will travel with some tender seedlings on steamships with palm-greased engines to South East Asia, to British and Dutch colonies where imperial powers took advantage of the social, economic, and environmental disjuncture they themselves had unleashed to open up new lands for palm oil plantations, and to recruit dispossessed and migrant workers, often through techniques of debt bondage.[17] Today the independent nations of Malaysia and Indonesia are undisputed palm oil superpowers, though the legacies of colonialism remain pivotal in the industry. We will follow the oil as it seeps through the fabric of our world, becoming the fat of the world's poor and the grease of capital's global empire.

This is a story of human sacrifices: the sacrifice of people and places made cheap by a system driven by profit, a system that seduces most of us in one way or another, as consumers, as entrepreneurs, as people just trying to survive. By following palm oil, I want us to recognize the way we are all bound up in a global paradigm none of us chose, but that benefits some vastly more than others, and places so many on the altar of accumulation. Primarily, those sacrificed are the exploited workers who cultivate and process the commodity, and those who have been dispossessed of their relation to the land by its reckless spread, driven by the desires of the market of which we are all an unequal part. In the corruption-prone palm oil industry, where local elites, large corporations, national governments, and international agencies dance together, labor abuses are rampant.[18] Millions of animals and whole species are snuffed out as rainforests burn to open new plantation lands. Under blood red skies, the carbon released tears at the lungs of workers, villagers, and Indigenous people who live and labor in the shadow of the palm industry and smog suffuses the skies over South East Asia. But the deforestation and the carbon released by burning also represent one of the gravest threats to the global ecosystem. Also sacrificed is the health of millions around the world who this same set of systems has rendered so poor that palm oil and its cheap derivatives become staples of their diet, with devastating consequences.

This book tells something of a chronological story, but it is not a history. It is an attempt to trace the contours of something hidden in plain sight. It has its origins in a series of modules I developed while teaching material culture and capitalism at the Nova Scotia College of Art and Design. I was seeking to do more than sensitize these fine art students to the facts and figures about this substance, which was found in the paints, inks, dyes, resins, computers, plastics, and other materials with which they worked. I also wanted to explore with them the dense enfolding of past and present, here and there, us and them. As a result, our story here is impressionistic and idiosyncratic. It is written by someone who has never seen a palm tree, never visited a plantation, by an end user of a commodity, trying to find his way back to the source and, through that journey, better understand the world.

This book will not offer a comprehensive overview to the palm oil industry; I encourage you to consult journalist Jocelyn C. Zuckerman's *Planet Palm: How Palm Oil Ended Up in Everything and Endangered the World*. Neither will it offer a systematic history; this has already been done with admirable skill and sophistication by Jonathan E. Robins in *Oil Palm: A Global History*. I am not going to catalogue the crimes of today's palm oil industry, or its cynical manipulation of the truth to hide those crimes behind aggressive public relations campaigns and toothless voluntary regulatory schemes. You will find no shortage of non-governmental organizations focusing on environmental and labor rights around the world, or news magazines who have published high-quality exposés and reports on the palm oil industry. Nor will I offer a multispecies ethnography or new materialist account of the nuanced relationship between *homo sapien sapiens* and *E. guineensis*.[19]

Instead, in *Palm Oil: The Grease of Empire*, I am writing to you inspired by artist Simryn Gill and anthropologist Michael Taussig's invitation to consider the question: what does it mean to be human in a world made of palm oil, in the sense that it is in or on or part of the production of so many things we use every day? What does it mean to tell our story to ourselves in bodies that are made of palm oil, in the sense that we, all of us, metabolize it from our food or wear it on our skin?[20]

This book is in part a friendly critique of journalists, environmentalists, and human rights campaigners who tell the story of palm oil as one of poor, benighted, abused workers, farmers, peasants, Indigenous people, orangutans, and rainforests "over there," in palm oil producing regions, who need enlightened consumers "over here," in palm oil consuming regions, to wake up and make different consumer choices.[21] As useful as these narratives may be as a way to raise funds, mobilize boycotts, or orchestrate political campaigns, this approach retains a model of charity based on longstanding colonial and white supremacist tropes. This discourse has proven to be easily captured, coopted, or subverted by the advocates of the industry being critiqued. Today, the palm oil industry and its friends in many governments and big non-governmental organizations (NGOs) have quite successfully wielded liberal environmental and human rights discourses to further their agenda of extraction and profit.[22]

At stake for me is not so much which campaigns will work to arrest the worst excesses of the palm oil industry. This goal is indeed important: millions of lives are at stake, and whole ecosystems hang in the balance. But my ambition is more tangential. Against the consumerist "we" that is all too often asked to act on behalf of workers, orangutans, rainforests, or "the climate," I wonder if another "we" is possible that sees all of these things as part of the same web of mutuality. My concern is if, in telling a more complex, expansive, and experimental story of palm oil, we can come to recognize ourselves, individually and collectively, anew. If we, a global species, are all now made of palm oil, and if the world is made of palm oil, perhaps we need new narratives for understanding ourselves if we are to come to terms with what we might do to better transform the world that is always transforming us.

Similar to the way it occurs in the web of global commodities as a universal but invisible substance, in this book palm oil appears in the in-between of history, culture, politics, and economics. Ours is an oily narrative: themes, timelines, and arguments slide around, seep into and stain one another.

We know from the anthropological and historical record that a vast array of civilizations have practiced human sacrifice, and for a huge number the market is taken to be a natural, inevitable, and apolitical entity, governed by eternal but discoverable laws.[24] But the market is a human creation, our own creative and cooperative powers reflected back to us in a distorted mirror. It is in no small part an unconscious creation of our imaginations, held in place not by some terrifying overlord but by its own relentless, self-reproducing logic. As we each try and survive and perhaps succeed in a market-driven capitalist society, we add weight to its momentum. Yet like most humans who make sense of the world from within a particular cosmology (and perhaps more than many), we are largely unaware of its influence.[25] As Sylvia Wynter argues, contemporary capitalism's cosmology emerged from a white supremacist worldview that took shape through and helped facilitate colonialism and the slave trade, one based on the supremacy of the figure of the self-actualizing, independent actor, which lately takes the form of *homo economicus*, the neoliberal subject.[26] This imaginary template is presented as both the paragon of humanity and reflective of the ultimate truth of human nature: rational, competitive, acquisitive. This imaginary ideal was forged in the crucible of empire and is coherent only in contrast to its many "others." In particular, the cosmology of capitalism and *homo economicus* emerges by denigrating those non-European "genres" of humanity as, at best, self-mystifying imitations that had to be "civilized" out of existence or, at worst, the realm of subhumans, fit for nothing more than exploitation or elimination. Particular cosmologies justify, normalize, and render "necessary" specific forms of human sacrifice. Capitalism's order of human sacrifice is in some ways unique and in some ways fairly conventional.

To tell a story of palm oil as capitalist human sacrifice, we begin with the 1897 British punitive expedition against the Edo Kingdom in what is, today, Nigeria, allegedly pursued as a civilizing mission to avenge an affront to British honor and intercede and eliminate an African kingdom whose practices of human sacrifice had been salaciously broadcast in the British press.[27] In fact, a key motivation—beyond the sheer lust for conquest and the machinations of petty officials—was to secure West Africa's lucrative palm oil production capacity for British corporations.

Our story ends in the profit-driven American prison system, where incarcerated people are fed diets of cheapened foods derived from palm oil, a disproportionate number of whom are of African descent. Most have been criminalized as the result of actions stemming from racist life circumstances of systemic poverty, trauma, or social neglect. Here, ramen noodles, "soup" as they are called inside, has become a currency as inmates struggle to survive a system of mass capitalist human sacrifice.[28]

In between we will learn about the way the modern palm oil industry arose from the ashes of the slave trade, and of the way palm oil was the secret ingredient and lubricant of the industrial revolution. We will learn how palm oil, in the hands of European investors and corporations, became integral to colonial warfare and to the rise of modern art and the institution of the imperialist museum. We will, of course, dwell with the conditions under which palm oil today is produced on or in the shadow of plantations in Indonesia and Malaysia and beyond, and the humanitarian and ecological catastrophes they have unleashed. We will also see how the response of NGOs and governments from the Global North have failed to address the situation because they rest on flawed assumptions about how this system of human sacrifice works. And we will concern ourselves with the way that palm oil had become the fat of the world's poor, the trace, in all our bodies, of an age of human sacrifice.

This book's conclusion meditates on what it might mean to move beyond the framework of consumer activism and, instead, recognize in palm oil our own reflection. We are a species uniquely capable of transforming ourselves as we transform the world. The transformation of the palm oil industry will be led by those most affected by it, but it will be part of a worldwide rebellion against the cosmology of the market and its confinement of our imaginations. This necessarily implies we reimagine what it means to cooperate and become plurally human.

# Whose punishment?

| | |
|---|---|
| **Killmonger**: | Good morning. |
| **Curator**: | How can I help you? |
| **Killmonger**: | I was just checking out these artifacts. They tell me you're the expert. |
| **Curator**: | Ah . . . You could say that. |
| **Killmonger**: | They're beautiful. Where's this one from? |
| **Curator**: | The Bobo Ashanti tribe . . . present day Ghana . . . nineteenth century. |
| **Killmonger**: | For real? What about this one? |
| **Curator**: | That one's from the Edo people of Benin . . . sixteenth century. |
| **Killmonger**: | Now, tell me about this one. |
| **Curator**: | Also from Benin, seventh century. Fula tribe, I believe. |
| **Killmonger**: | Nah. |
| **Curator**: | I beg your pardon? |
| **Killmonger**: | It was taken by British soldiers in Benin, but it's from Wakanda. And it's made out of vibranium. Don't trip. I'mma take it off your hands for you. |
| **Curator**: | These items aren't for sale. |

| **Killmonger**: | How do you think your ancestors got these? You think they paid a fair price? Or did they take it, like they took everything else? |
| **Curator**: | Sir, I'm going to have to ask you to leave. |
| **Killmonger**: | You got all this security in here watching me ever since I walked in. But you ain't checking for what you put in your body. |

So we were introduced to the villain Erik Killmonger in *Black Panther*, the fourth highest grossing film of all time.[29] In this scene, Killmonger and his accomplices poison the curator of the fictional Museum of Great Britain as a means to create a brief distraction that allows armed men to enter the museum under the guise of paramedics and seize a vibranium artifact. The fictional metal originated from the site of an ancient asteroid impact in the imagined composite African kingdom of Wakanda and has profound metallurgical, energetic, magical, and military powers. Wakanda has leveraged its monopoly on vibranium to develop a highly technologically advanced society, but uses that technology to hide itself from the world, masquerading as a very poor nation for fear that foreign lust for vibranium would trigger war.

Is vibranium a metonym for the rich mineral, cultural and human wealth stolen from Africa in the centuries of the transatlantic slave trade, imperialism and, later, neocolonial extractivism? One can recognize in its story the shadow of palm oil: a likewise almost magical transformative substance, capable of being transmuted into food, medicine, technology and, as we shall see, weapons. *Black Panther* offers an Afrofuturist vision of an alter modernity that might have been had African polities been allowed to pursue their own autonomous patterns of technological and political change.

The popularity of the #TeamKillmonger hashtag throughout *Black Panther*'s theatrical run and after revealed the groundswell of sympathy for his tragic character. As the film develops, it is revealed that Killmonger not only wants revenge on Wakanda for abandoning him, the son of the king's exiled brother, he also wants to take over the Kingdom and puts its fearsome vibranium arsenal in the hands of the wretched of the earth.

Killmonger's heist takes place in the "West African Exhibit" of the fictional Museum of Great Britain, a set that bears an unmistakable resemblance to the Sainsbury African Gallery of the real-world British Museum, which exhibits part of the world's single largest collection of the famous Benin Bronzes of the Edo Kingdom in what is today Nigeria. These remarkable ivory, copper, brass, and bronze statues, wall plaques, and other artifacts were looted at the conclusion of Britain's 1897 Benin Punitive Expedition, which also resulted in the destruction of the Edo Kingdom and its world-famous capital city, as well as the deaths of uncounted thousands of its people.[30] Some of the fictional artifacts Killmonger quizzes the curator about are pieces that bear a striking resemblance to the real-world Bronzes. The Wakandan vibranium artifact that is the target of his reclamation (or theft) was, he tells the curator, "taken by British soldiers in Benin."

The Edo Kingdom was known to Europeans since the fourteenth century as a well-organized and powerful West African polity, organized under the entwined spiritual and political leadership of the Oba and his royal house, notably the powerful dowager queen. The splendor of its capital city, with its sophisticated earthen-work ramparts, was legendary and, in the seventeenth and eighteenth centuries, rivaled the largest European metropoles in size and social organization. In this highly stratified society, the artifacts known collectively today as the Benin Bronzes represent centuries of Edo history and their cultural importance has no exact European analogue. We have a handful of photographs of the artifacts in situ, including palm oil polished busts sitting on altars, burnished wall plates decorating thick pillars holding up the wide rooves of public buildings and naturalistic statuary decorating exterior architecture. Some of the artifacts were ornamental, representing the fruit of generations of specialized craftsmanship and a specialized division of labor. Many objects served as ledgers of aristocratic lineages or the kingdom's historical events—a kind of sculptural library. Others were manufactured primarily for trade, including with Europeans, trade often denominated in or lubricated by palm oil. Some were spiritual, part of altars to ancestors and to gods of the Edo royal house, altars that would see devotees use palm oil in hundreds of different sacramental ways, including human sacrifice.[31]

The Bronzes, represented in some senses the artistic crystallization of an African empire built on the wealth of palm oil. Its capital was the economic, administrative, cultural and religious hub of an expansive system whose lifeblood was that substance with so many uses.[32] The conditions and relations of palm oil's production might depend on one's status within the empire: wealthy gentry families might oversee large workforces of slaves who tended the plants, climbed them to harvest their fruits, oversaw the fermentation process, expelled and clarified the oil, and transported the commodity to market. Nations or communities enthralled to Edo paid tribute in palm oil too. Workers and other non-elites produced their own oil for consumption or to sell or to meet the demands of taxation. Like many of the diverse societies in the region, palm oil played a vital role, and the relations of its production helped define a whole way of life.

Today, thanks to decades of efforts by the Nigerian government, the descendants of the Edo royal court, and dedicated allies, there is a strong movement to compel museums and other cultural institutions to repatriate these artifacts.[33] The Bronzes were seized indiscriminately as the capital of the Benin Empire was put to the torch. They were auctioned off in Europe later that year by the British to defray the costs of the expedition itself. Today, ornamental and decorative items, as well as deeply charged ceremonial and religious objects, sit, helter-skelter, in public and private collections in London, Berlin, New York, and elsewhere. More often than not, they are displayed as part of ethnological collections that implicitly or explicitly glorify European culture.[34] As anthropologist J. Lorand Matory notes, in European museums African spirited objects are fetishized anew, made to represent both the barbarism of that Other culture that originally created them as well as the benevolence of that civilization that collects and "protects" them. They are commandeered to serve as props in the melodrama of a universal history where European culture and institutions play the heroic lead.[35] The movements that are rising today to demand the restitution of these and other artifacts, or the tearing down of the statues of their kidnappers, aim not only at correcting history's wrongs but undermining the white supremacist cosmology that empowered their theft. That cosmology continues to rule, not through the formal trapping of nineteenth century imperialism, but through the more subtle but no less lethal neocolonial frameworks of the market.

In the British Museum, the Sainsbury Wing has incarcerated the Benin Bronzes since 1991. It is, not incidentally, in the basement—the crypt as it were—as if the spirits the objects emobdy represent, or the violence of their theft, need to be safely sequestered underground. They are located almost exactly under the famous round reading room that used to house the British Library, the Empire's preeminent public collection of statistical economic records. It was here that, famously, Karl Marx researched and wrote much of *Kapital*, where Friedrich Hayek made notes for his *The Road to Serfdom*, and it is mere steps from John Maynard

Keynes's Bloomsbury home. The wing itself bears the name of the wealthy Sainsbury family of grocers, many of them peers who sit in the United Kingdom's House of Lords. The fortune which enabled their gift to the British Museum derives from their successful supermarket chain, where half the items on the shelves contain palm oil.

The British Punitive Expedition against the Edo Kingdom, as the title baldly asserts, was a barely authorized mission of racist imperial vengeance.[36] By the 1840s Britain was already seeking to consolidate control over territories today known as Nigeria as its exclusive zone of influence. This was consummated and recognized in 1885 at the notorious Berlin Conference, where European colonial powers carved up Africa over a large oaken table in an ornate room in the cold, wet winter of the capital of the newly unified German Empire. The conference was preceded by decades of jockeying between European powers to secure (usually highly prejudicial) exclusive treaties with African monarchs and trading cartels. Following the Berlin Conference, the British Empire undertook "punitive expeditions" with regularity to punish African kingdoms and cartels for refusing to abide by these extortionate treaties, for stopping palm oil exports, or for in almost any other way inconveniencing the accumulation of British capital.

The expedition was an example of the kind of imperial action that brought to a close a tectonic century in the region. The abolition of slavery in the British Empire began with restrictions on the trade of enslaved people in 1807 thanks to a worldwide rebellion, first and foremost by enslaved people themselves, but also by the working classes, abolitionists, and religious reformers in England. In the wake of this people's victory, many of Liverpool's merchants, whose fortunes had been made through this heinous atrocity, turned their ventures toward the extraction of palm oil from West Africa to Europe, including in the territories then known to the British as the Oil Rivers, not for petroleum (that would only come to be important in the following century) but for palm oil. The transoceanic market for West African palm oil emerged within the slave trade, when it was often taken on board to feed enslaved Africans on the middle passage, or to grease their bodies at the end of the voyage to add marketable luster or hide the severity of wounds and scars from prospective buyers.[37]

Meanwhile, palm oil became one of the essential ingredients of capitalist, imperial modernity. To give a sense of the scale, palm oil imports to Britain (which controlled around three quarters of the global trade) increased by a factor of 566 from 1807 until its peak in 1897.[38] In that year, the 63,147 tons of refined oil that was offloaded in Liverpool, London and Bristol represented somewhere between 8–27 million workdays of African labor, which includes growing, harvesting, processing, transporting, and vending the oil, as well as of enforcing the conditions of exploitation and unfreedom that made the rest possible.[39] Statistics from 1830 indicate that palm oil sold at Liverpool markets at ten times the price for which it was purchased from African producers.[40] In Europe, palm oil greased the wheels and engines of empire, providing a crucial lubricant for industrial machines, railway locomotives, steamship engines, and more.[41] It was also crucial for a number of new consumer products of the late nineteenth century: soap, cheap candles, tin cans, and later, edible margarine. It was lust for this almost magical substance that brought Europeans to West Africa. The popularity of ornamental palm trees in Victorian London testified to the importance of the crop as both a raw material and a trophy of Empire.[42]

In the early part of the nineteenth century, British traders were content to simply buy African palm oil from the decks of their ships or from hulks docked at the mouths of rivers, allowing coastal African elites to do the dirty work of extracting the unguent from the hinterland. Vast fortunes were made as European merchants issued trade goods to African traders on credit, or "trust," to secure steady supply. By the mid-nineteenth-century older British trading family businesses were superseded at coastal entrepôts by "palm oil ruffians," unscrupulous and often extremely violent European entrepreneurs who were reviled at home for their crude dispositions, but valued for the profits they returned to British investors. As European demand glutted the market and led to a decrease in prices after the 1850s, as steamships, antimalarials, and other technologies emerged, and as joint stock corporations began to manage the trade, Europeans increasingly began to move onshore and take direct control of palm oil production. As Walter Rodney asserted, through the palm oil industry "Liverpool firms were no longer exploiting Africa by removing its labor physically to another part of the world. Instead, they were exploiting the labor and raw materials of Africa inside Africa."[43] Literally and figuratively, the same ship holds and financial ledgers that once distilled African lives into raw, disposable, fungible Black labor for sale in the colonial settler states of the Americas now transported the product of African labor to Europe.[44]

But by the end of the century, the demands of industry and rising imperial competition encouraged the British administration to take a firmer hand. While cloaked within a language of bringing civilization to a benighted people, European empires were, at this time, convinced that Africans were squandering precious resources because of their insistence on traditional forms of agriculture and commerce, which ought to be more rationally and scientifically organized, allegedly for Africans' own good.[45]

It is important to note the centrality in this story of the capitalist structure of the corporation: that strange legal fiction made real, the ultimate fetish. Born to facilitate the risky colonial and later slave-taking ventures of the rising European bourgeoisie, this strange, monstrous entity, which exists purely to generate profit for shareholders, was afforded legal personhood long before most of the world's inhabitants were recognized as fully human by European law. By the time of the Punitive Expedition, the palm oil trade was dominated by (for the time) large corporate interests capable of not only generating significant capital and retaining talented traders and managers but also hiring private military forces and powerful lobbyists. Shares in these companies were part of the ebb and flow of investment in Liverpool and London, where a rising capitalist class benefitted without even having to concern themselves with the conditions under which the oil that was the source of their wealth was produced.

Meanwhile, the 1897 Punitive Expedition was justified and sold to the British public in terms of the inherent civilizing influence of free trade and as a means to end the barbaric practice of human sacrifice.[46] For some decades the Edo Kingdom had been painted in the British press as the City of Blood, ruled by a merciless tyrant who demanded ceremonially oiled flesh to appease his cruel gods.[47] American Missionary Robert Hamill Nassau efficiently expressed the common Eurocentric demonization of West African political and spiritual practices when he described them as "governments of the fetish," sustained by the cynical and tyrannical manipulation of subalterns by bogus supernatural threats.[48] But perhaps even more egregious for the British than the human sacrifices was the way this supernatural "juju" power was cited by the Oba as a reason to prohibit or monopolize the exchange of certain commodities, imperiling the fetishized European notion of "free trade."[49] Palm oil, for one, was considered a holy substance and therefore under the jurisdiction of the Oba, who strictly regulated its trade to the chagrin of British merchants.[50]

How, why, and with what consequences the Edo Kingdom practiced human sacrifice remains a matter of historical debate, one clouded by the fact that almost the only "reliable" records we have are those reported by Europeans, often for the purposes of defamation and to legitimate missionary or military intervention.[51] While only a handful of historians doubt that these practices occurred, some argue that, rather than some gory religious rite, these executions had more in common with capital punishment, though within a highly religious society where acts of state were freighted with spiritual significance.[52] Others argue that these human sacrifices were simply a ceremonial expression of what *all* empires do: murder their enemies, use lethal violence to maintain order, suppress vassal peoples who might otherwise become rivals, and then dress it all up in the language of cosmic necessity.[53] Anthropologist William Pietz suggests that the fields of crucified corpses reported by members of the Benin Punitive Expedition as they made their way to the capital in 1897 were probably not the norm of that society but the desperate acts of religious elites in the face of what they saw—correctly as it turned out—as an existential threat to their civilization: if the Gods that had lifted Edo to greatness had abandoned the kingdom, or had starved for lack of appropriate devotion, mass sacrifice may have been a desperate last-ditch effort to reawaken their favor or revive their vitality.[54]

Perhaps such orgies of blood happen at the end of all empires when, as Tvetan Todorov argues, not only myths but entire systems of symbolism and meaning-making crumble.[55] Perhaps it explains the vindictive cruelty of today's sacrificial global capitalist economy, where millions find themselves on the altar of a market that seems not to hear our prayers or honor our sacrifices?

We may never know conclusively what role human sacrifice played in the Edo Kingdom, and it doesn't much matter for our story. What matters more is the *uses* to which the British put that story of human sacrifice. It was in the name of intervening to stop this heinous practice that they justified their invasion.[56] And yet the spectacularized stories of sacrifice helped also mystify, normalize, and diminish the sacrificial violence of that very invasion, and of the system of imperialism of which it was a part. It left entire civilisations in ruins, whole populations as vassals. In focusing their attention on the Edo Kingdom's explicit, sensational acts of religious bloodletting, the British were able to ignore the fact that their Empire, equally and with more dire consequences, was placing human beings on the altar of white supremacy and capitalist "free trade."

Africans in the ruins of the Edo Kingdom and elsewhere fought back. Displaced nobles and soldiers of the Oba waged guerilla war against the British occupation for two years after the fall of the Kingdom.[57] The Oba himself, though in ignominious exile, never formally conceded his throne and today his descendants still occupy that seat in a ceremonial capacity in Nigeria's Edo province. Even where military and political resistance was made impossible, African people found numerous ways to resist and rebel.[58] These included refusal to work, the seemingly accidental breaking of infrastructure or spoiling of supply, the siphoning of resources or even smaller acts.[59] The 1929

"Women's War," whose protagonists numbered over 10,000, erupted because of the way the British colonial administration ruled through the nomination of typically corrupt male Warrant Chiefs who collected extortionate imperial taxes, notably on palm oil merchants.[60]

The story of palm oil's extraction in this period is not a straightforward one of innocent and unified African people withstanding evil European incursions. The imposition of capitalist processes intersected and entangled with local hierarchies and injustices, including the persistence of slavery and the dominance of both longstanding and emerging power elites.[61] But the fact of this complexity ought not discourage us from paying attention to the way capitalism, racism, and colonialism articulated one another in this context, and persist to our present age.

In her fascinating study of the dense networks of entanglement, contradiction, and conflict in Indonesia's forests, many of which are today being razed for palm oil plantations, anthropologist Anna Tsing encourages us to recognize that capitalism advances through the friction between its universal logic of accumulation and the particular cultures, lifeways, and structures of power in each locale.[62] Contrary to the dreams of a "friction-free" capitalism promised by billionaire, philanthropist, and self-styled intellectual Bill Gates, capitalism is defined by conflict and tension.[63] Gates is only the latest of a long line of capitalist thinkers to dream that the free market, if allowed to flourish without regulation, will create a smooth world where the hardworking and talented, no matter their origins or station in life, could compete to succeed, and that this competitive striving would have beneficial effects for global society at large: greater wealth and greater innovation. Even many critics of capitalism fall prey to the myth of frictionlessness in their attempts to explain the nefarious clockwork of this demonic machine that metabolizes people and the earth not only into profit, but into the energy that fuels system's endless reproduction and expansion. Tsing's insistence that we look to the friction encourages us to recognize that what moves this machine forward (and halts its "progress") is the friction that stems from forms of complicity and resistance, acquiescence and refusal that define every point where the abstract system of capitalism encounters the material realities of the entangled earth and its people.

If we follow Tsing's powerful metaphor, we might also ask: how does capitalism overcome friction? Here, we might look to palm oil, the grease of empire. In different eras, this substance has played a role in helping that system facilitate its expansion in a multitude of ways. By following palm oil, then, we can bring to the foreground the complexity of power and resistance that is the stuff of the world we make together and some "we" can reacquaint itself with how the past and present are enmeshed.

# Whose fetish?

The primary reason the British and other European powers were interested in West Africa's palm oil production capacity was their need for industrial lubricants. The rapid industrialization of Europe in the nineteenth century demanded millions of liters of refined palm oil annually, either on its own or in combination with other oils, as a means to cool and maintain industrial apparatuses in factories, public works, railways and more. It is likely not an exaggeration to say that the literal and figurative wheels of empire, from locomotives to steam engines, were greased with refined West African palm oil.

It was primarily toward this market that the Liverpool merchants turned after 1807 when denied their massive profits from the slave trade.[64] They commissioned chemists and investors to develop new formulae and applications for palm oil and to promote its use among industrialists and others.[65] Prior to the widespread adoption of palm oil, Europeans typically used highly refined tallow from cattle and mutton. But by the nineteenth century, with the increased demand for animal fat from European imperial militaries and the interruptions of supply chains by war, the cost rose. In addition to being relatively expensive, tallow could be smelly and turn rancid, especially on the long sea voyages and in the hot and humid jungles of empire. While the initial costs of importing palm oil were considerable, an economy of scale and the advent of the steam ship soon saw prices fall.[66] Indeed, by the time of the Benin Punitive Expedition of 1897 the price of palm oil had dropped precipitously, thanks in part to the proportion of West Africa now dedicated to its manufacture and competition between European merchants.[67] Also, by this time technology for extracting fine oils from coal and petroleum were increasingly sought-after for their price and the precision with which they might be engineered for specific industrial applications.

One might, then, expect that the British empire would find the commodity too cheap to justify the Benin Punitive Expedition's investment of military force and, later, colonial administration. Indeed, this initially led to a great deal of hesitancy in the colonial office who did not wish to be encumbered with another unprofitable and rebellious "protectorate."[68] But as Marx instructed and Rosa Luxemburg affirmed, contrary to the rosy predictions of bourgeois political economists, capital's response to falling prices and saturated markets is simply greater violence to find new vistas of accumulation. Empires compete to open new markets through imperialism, or go to war to annihilate people, or destroy over-accumulated wealth.[69] In the case of the West African palm oil zone, this took the form of European corporations expanding their coastal operations inland in an attempt to cut out African middle-men and undermine the power of local elites who controlled the supply of the commodity. These capitalists demanded that their respective empires use military threats or procedures to enforce these conditions as needed.[70]

Throughout the late nineteenth century palm oil was the most popular fuel for producing candles in the period of the decline of the whaling industry and prior to the ascension of coal and petroleum-based oils and the advent of widespread electrification. Unlike tallow, palm oil candles burned cleaner and with less odor, and were cheaper to produce, largely because the vast bulk of the labor of extracting the oil was "offshored" to hyper-exploited Africans.[71] The cheap accessibility of candles as a method of illumination in Europe was part of the sacrifice of working class time on the altar of accumulation. While many factories and public buildings may have had gas lighting, in the tenements of the poor candles still played a major role in providing both heat and light, notably for the ragpickers, ropemakers, seamstresses, and other pieceworkers whose intermediate, handmade products fed into the industrial apparatuses of the metropolis. Candle sales only began to diminish in the 1890s.[72] Here, palm oil became a weapon in the conquest of what Jacques Rancière alludes to as the "proletarian night,"[73] the dark hours, on the shadow side of modernity, when the eros of the radical imagination flourished.[74] The cheap candle here was not so much an evil capitalist machination as it was a tragic tool working people used to entrepreneurially commodify their own time under conditions of profound economic duress.

Palm oil candles were presented to European consumers as an ethical choice. The Liverpool palm oil merchants and their clients, notably candle and soap makers, advertised their products as a boon to West Africans that would encourage them to pursue palm oil export as a way to transition away from the institutions of slavery.[75] In what some historians describe as one of the first modern marketing campaigns, Prices' Candles broadcast the idea that the emerging European consumer could change the world with their pocketbook. The hard work of cultivating *E. guineensis,* harvesting its fruits from high up in the canopy, fermenting the seedpods then crushing, boiling and refining their oil, which was often undertaken by whole extended families conscripted by debt bondage to the task, was imagined by Europeans to be honest work for Christian souls. This, in contrast to what they imagined was the superstitious, idle and tyrannical state that had existed before, and held in place by slavery's need for compliant local chieftains. European consumers were told that burning palm oil candles was doing God's work by creating a market for a product that encouraged peace and spiritual and material uplift.[76] Such marketing campaigns adopted and repurposed ideas, tropes, and imagery from the abolitionist struggle, offering later generations the opportunity to participate in a fondly remembered moment of ethical righteousness. But whereas that earlier great upheaval was based, at least in part, on British proletarians' sympathy with people across the world likewise exploited by capital and made disposable by empire, the advertising campaigns of candle and soap manufacturers traded in a growing "commodity racism" that capitalized on the alignment of British class interests around narratives of shared imperial supremacy and white civilization.[77] This is one way in which the dangerous bonds of the transnational solidarity of exploited workers were increasingly replaced by condescending notions of consumer charity.

In the mid 1830s, chemists developed techniques that led to the bleaching and deodorizing of palm oil for use as a base for soap. British consumption of soap doubled between 1801 and 1833.[78] While it began as an additive to luxury soaps, the industrial revolution transformed palm oil into the basis of what became an everyday necessity.[79] Companies like Lever—predecessor of today's Unilever, still one of the world's single largest consumers of refined palm oil—were heralded as beneficent capitalist visionaries for their enlightened policies toward their European workers. This was emblematized by the paternalistic utopian scheme of William Hesketh Lever's factory town, Port Sunlight, upriver from Liverpool, a region that since the mid-1800s had outpaced London as the empire's leading producer of soap.[80] There, workers who might otherwise be forced to live crammed into festering tenements in the thick smog of the industrial city were invited to inhabit generous (for the time) townhouses in a purpose built estate, a place where vices like drinking, non-matrimonial sex, and gambling were strictly policed.

But Lever's treatment of African workers directly inherited the forms of discipline and cruelty forged on these same coasts in the slave trade, renovated and combined with the latest theories and practices of plantation management. This was especially the case in Congo where Lever gained major concessions from Belgium to establish an intensive palm plantation, applying industrial methods to the growth, cultivation, and processing of a fruit that, until that point, had still allowed smallholders to maintain a bit of autonomy and skill.[81]

The plantation model, as oil palm historian Jonathan E. Robins points out, is a profound process of creating a "new nature" and "'delocalizing' a space by clearing vegetation; evicting human and animal inhabitants; and contouring, draining, and irrigating the landscape until it fits a universal model." It only occurs within a broader economic landscape of input resources and markets for outputs.[82] The plantation, as a uniquely modern and fundamentally colonial technology, transmutes entangled land and labor into private wealth. As Jason Moore and Raj Patel argue, the plantation represents the paradigmatic technology by which capitalism "cheapens" land, life, food, care, and fuel.[83] It is not without reason that theorists have recently encouraged us to recognize our epoch as the plantationocene in recognition of the deep—indeed geological—transformations of the planet wrought by this invention.[84]

Lever's West African plantations, like his experiment at Port Sunlight, were never the success he hoped, thanks in part to the everyday forms of resistance and refusal practiced by workers.[85] In the Belgian-controlled Congo and other places Lever obtained concessions, workers routinely appeared to ignore or intentionally misunderstand instructions, found ways to siphon resources, or refuse to work or work hard, causing the company to "innovate" ever more draconian and extortionate labor relations, including two things that proved and continue to prove today devastatingly effective: debt bondage and extortionate contracts and the explicit or implicit use of paramilitary and gang violence.

Though many today see soap as an indispensable part of daily life, in the nineteenth century a market for soap needed to be created in Western Europe through aggressive advertising campaigns.[86] As palm oil made the industrial production of soap cheap, its manufacturers were among the first and most enthusiastic to use the then-new media of picturesque magazine and billboard advertisements to create a want and then a need.[87] Though a relative latecomer to the industry, Lever's success stemmed less from the quality of his product or the efficiency of his operations and more from his acumen as a marketer. By the 1920s, palm oil importers including Jurgens and Lever were both so rich and so keen to expand consumer markets they acquired high street retail stores, the predecessors of today's supermarkets, to gain immediate access to consumers, including the aptly named "Home and Colony" chain.[88] By the end of the century, Britons on average consumed some 17 pounds of soap per year, with a tropical fat content of at least 42%.[89]

This was done through associating soap with the dominant values of the age. Soap was advertised first to middle class and then to working class women as the means to care for and protect the family and the home from an outside world of danger, filth, and contamination. Such advertising both capitalized on and helped reproduce changing notions of gender and the family, notions that, as radical theorist Silvia Federici makes clear, ultimately served capitalist accumulation.[90]

It was thanks in part to such marketing that the notions of the modern proletarian woman was normalized: an unpaid domestic worker expected to provide reproductive labor that enabled wage-earners to return to work each day and to gestate and nurture a new generation of workers, too. While European women also worked in the formal capitalist economy in factories, in domestic service, and elsewhere, they were legally and customarily paid far less, making them reliant on male wages of fathers, husbands, or other kin.

Gender norms, including those promoted in soap advertisements, helped normalize a world in which reproductive labor was assumed to be the natural inclination, the highest priority, and the greatest desire of women, associated with cleanliness, morality, and the possibility of a good life.[91] These aligned with bourgeois values which stressed the patriarchal family as the seat of both economic, moral, and political life.

Soap advertisements depicted women bathing infants, or bucolic scenes of children playing in innocence and peace with devoted mothers and maids looking on fondly. At the same time, cities like London, Manchester, Birmingham, and Sheffield were suffocating in the soot from factories and tenements, deprived of proper municipal and household sanitation. This was largely due to the refusal of the wealthy, who controlled the levers of government, to take seriously the public health risks that, in any case, rarely reached their suburban or country homes.[92] Workers, both adults and children, routinely came home from grueling 12–14 hour factory shifts slick with sweat, grease, and grime. In the interests of maximizing rent landlords had little incentive to offer working-class families large or well-spaced living spaces, let alone luxuries like gardens or reliable running water.

When industrially produced soap became cheap enough for the working classes to buy, it was promoted as a tool for taking personal responsibility for hygiene and health in a world where that was largely impossible. Public hazards that were the result of vast and lethal inequalities were presented as private risks to be managed through consumerism and changes in personal behavior. Reformers distributed condescending pamphlets teaching the urban poor how to use soap for hygiene and the moral uplift of the race.[93] For the middle class, soap was presented as the magical means to protect and preserve the bourgeois home from a world of economic, political, and moral threats that, ironically, almost all stemmed from the bourgeois world system and capitalist economy that made the patriarchal middle-class home possible in the first place. For the working classes, the imperative to buy and use soap promised a means to mitigate the risk of noxious filth and disease that pervaded late-Victorian society thanks to forces well beyond their control. In many if not most societies where human sacrifice is practiced, ritualistic ablution of the victim is prerequisite.[94]

As with candles, palm oil derived soap was advertised using tropes of a white supremacist empire.[95] In some, Black children magically became white when soap is applied, associating darker skin with animalistic filth and degeneracy and whiteness with purity, hygiene, and civilization. Later, advertisements were to explicitly propose that the bringing of soap and "modern" hygiene to the "lesser races" of the earth was part of the mission of Empire and the White Man's Burden. An imperial and white supremacist project that had largely benefitted the ruling class was sold to first the middle then the working classes as a shared racial project, the better to ease mounting class tensions.

In this sense, soap represented a profoundly "fetishized" object. For Karl Marx, who doubtless saw such advertisements almost daily, as well as the squalor of working-class life near his home in London's Soho, commodity fetishism was ultimately a process of forgetting.[96] It describes the way we come to see certain commodities as almost magical, with innate power. Soap, for instance, was sold as a kind of talisman to ward off the evils and risks of late Victorian society, to protect the home, the family, and normative gender roles, and to help establish and police the color line that was said to separate civilization from barbarism. But for Marx, fetishism was also a much more common and subtle process, whereby the fetishized object appears to us as if from nowhere, detached from the socioeconomic processes that produced it. Here, the origins of soap in the exploitation of African labor through imperialism is forgotten. So too is the labor of the working-class coopers, seamen, dockers, and transportation workers who conveyed it to European factories, and so too is the labor of those factory workers who transmuted the palm oil into the soap commodity.

But each commodity, if seen properly, is like a holographic shard of a greater capitalist totality, an image of all of the processes of exploitation that were invested in each stage of its production. Through the prism of the commodity we might glimpse all the extended socioeconomic relations on which those depend: the processes that produced the food that fed those workers, the exploitation of women in the home that reproduced those workers' labor power, the relations of direct or subcontracted violence and coercion that made each step of the process possible. Commodity fetishism is the everyday magic of unseeing this complex material reality of which we, each of us, is somehow a part and of forgetting the sacrificial violence on which the economy rests, hidden in plain sight.

Here, Marx's definition of the fetish dovetails with Sigmund Freud's, developed half a century later and seemingly without reference to the former. For the Viennese doctor, whose office was filled with the stolen spiritual objects of other civ-ilizations he obsessively collected, fetishism refers to a kind of psychic displacement. Here an object, a body part, a particular act, or a certain kind of person becomes the necessary gateway to ful-fillment.[97] Fetishism, while often harmless and actually quite common, is the result of some hitch in the development of the ego, usually in infancy or childhood, that has prevented the develop-ment of what Freud imagined to be normal psychic and sexual attachments. Underneath the fetish is typically some kind of secret, perhaps a trauma, of which even the fetishist is unaware, often something that cannot be admitted without fundamentally disturbing the fetishist's worldview or unified sense of self. Freud notes that fetishists often exhibit both a recognition and at the same time a refusal to recognize the artificiality and arbitrariness of their fetish: they know it is a proxy, and yet they still want it.[98]

Palm oil, then, especially when transmuted into candles or soap, was a nineteenth century European fetish. In the Marxist sense, these com-modities were taken for the magical means by which the gears of propriety, cleanliness, norma-tive gender, racial coherence, and social position were lubricated in the imperial capitalist machine. Like the grease in the actual machines of the industrial revolution, this sociocultural lubricant was largely forgotten. The fetishized palm oil com-modity hid within it the reflection of the entangled world of exploitation from which it emerged and of which each consumer is a part, which, if rec-ognized, might inspire more substantial forms of transnational solidarity. Freud's theory helps us to see how such commodities also lubricated the European sense of self which defined agency, desire, and value as a reflection of one's access to soap, candles, and other things, things which we both know and do not know originate from the sweated hands of other people.

The fetishization of soap and candles occurred within and helped to reproduce a kind of white supremacist capitalist cosmology, one that revolved around the figure of "economic man," *homo economicus*. Here, the white, male, European economic agent is taken as the universal template against which all other forms of life are measured and to which all people, or at least initially all European men, were expected to aspire. This is a figure who relates to the world of which it is a part through the framework of the market, the sphere of economic exchange. Europeans defamed the customs of the Edo Kingdom as fetishistic because they presumed that merciless strongmen were justifying their brutal power by insisting that control of certain sacred objects (like many of what would come to be known as the Benin Bronzes) endowed them with special powers, or that possession or exclusive use of those objects validated their sacred social rank and arbitrary authority.

Europeans accused Africans of mystifying their own cooperative relations by attributing magical powers to things they themselves made: ceremonial masks, rattles, ornaments, and other items. These things that a society itself produced from worldly matter and human labor were given supernatural authority, in a childlike form of

inevitable. We fetishize the market as an independent cosmos, when it is in fact an economy which our own actions help reproduce.

And yet both Marx and Freud's concept of the fetish were *themselves* fetishistic. As Matory argues, the European concept of the fetish erases its own material roots. The very *idea* of the fetish is itself a looted concept, derived from the way that European merchants, missionaries, and slavers sought to describe and defame African spiritual practices.[100] Like the now-common English words taboo (derived from the Tongan language) or totem (derived from Algonquian), the term was appropriated from its original context and made to work for European colonial thought, coming to signify the irrational obsessions of "primitive" people. Fetishism, in the mouth of the defender of colonialism, came to refer to the arbitrary sacred power assigned to certain objects or authoritative people which, Europeans surmised, was actually either a childlike common obsession or the cynical machinations of those who claimed spiritual authority to manipulate their gullible subjects. This meaning stood behind the accusation that the Edo Kingdom was exemplary of "government by fetish"—in other words, that the Kingdom had no form of political power beyond the arbitrary tyranny of their pious rulers who, for instance, used spiritual blather to demand human sacrifice.

What the weaponization of the word "fetish" does, then, is to accuse primitive others of infantile and dangerous obsessions, while the wielders of the weapon cast themselves as immune to such mystification.[101] *They* are fetishists, *we* never are.

Ethnological collections, like those that today house many of the stolen Benin Bronzes, reinforce this narrative. Yet in doing so the weaponized notion of the fetish cuts us off from a complex, introspective consideration of how we, collectively, make a world of things and how those things, in turn, co-create us each as individuals and as that "we" we imagine we are. It also permits us to "forget" not only the origins of things like palm oil, soap, or the concept of the fetish itself, it also encourages us to forget that we, each of us, are tangled up in those origins. We emerge from the same increasingly interconnected world that we are helping to remake, though under vastly different circumstances, with dramatically different levels of freedom and opportunity.

In the case of soap in the nineteenth century, the use of derogatory and racist imagery depicting benighted, filthy, fetishistic "savages" in need of European tutelage in hygiene draws on and reiterates this kind of weaponization of the notion of the fetish. At the same time, it makes soap a particular European fetish that becomes a kind of magical symbol for civilization itself. Meanwhile, it mystifies the actual origins of soap's key ingredient, palm oil, in the heinous violence unleashed in the African lands where Europeans first encountered the practices that they would come to call fetishistic. And yet this colonial violence, which I am identifying as sacrificial, beats like a telltale heart within the commodity and, like Freud's fetishized object, is both known and unknown, remembered and forgotten, banal and mystifying. Both the fetishization of such commodities and the set of assumptions tied up in the stolen concept of the fetish work to inhibit the radical imagination and funnel solidarity into a condescending form of charity.

This work of separation continues today in our world made of palm oil. It is not only that we "forget" that so many of today's everyday commodities, including soap, are made of or rely on this magical substance. Nor is it only that so many are ignorant of the massive human and environmental costs of that substance's production. This fetishism also transpires within the advocacy, activism, and NGO activity around palm oil today. For understandable tactical reasons, many campaigns take fetishistic targets, such as the plight of the charismatic orangutan deprived by deforestation of its habitat.[102] Other times they target specific products, such as the popular sugary spread Nutella. The artisanal palm oil producer, the picturesquely tragic Indigenous person, or the disenfranchised smallholder are frequently held up as the quintessential hapless victim of the Western consumer indifference or corporate greed.

This is in a way understandable: these organizations, whose staff typically have very sophisticated and nuanced understandings of the industry and its challenges, are competing in an intense attention economy to gain sympathy and mobilize action to try and affect a massive and powerful industry.[103]

And yet it contributes to a fetishism reminiscent of the soap advertisements of the late nineteenth century, and not only when it depicts brown-skinned people as in need of the benevolence of white consumers. Political economist Oliver Pye frames discourse of "sustainability" in the palm oil sector as a form of commodity fetishism. It seeks to mobilize consumer activism in order to demand better regulation that will improve the specific "management practice in some plantations," but which "has no impact on the regional dynamic of expansion." He continues that:

> Buying the commodity marketed as 'sustainable palm oil' might encourage a corporation to improve the management of their monoculture plantations, but it does not address the problem of the expanding area of monoculture plantations. Sustainability becomes a commodity, something that you can buy, rather than a social and political question.[104]

In fixating on this or that animal or human victim of the palm oil industry, or on this or that product or corporation, these campaigns inadvertently participate in a kind of forgetting of the broader totality that connects all aspects of the palm oil commodity: the early history of the human cultivation of oil palms in West Africa; the biotechnology used today to breed and select the cultivars, drawing in a world of universities and scientific journals, specialized scientific materials manufacturers and all the infrastructure they require; the steel, plastic, petrochemicals, and rubber used in the cars and trucks that take the harvested seedpods from the plantation to the processing plant; the food eaten by the workers in those plants; the forms of imperialism and neocolonialism that rendered the now-independent nations of Indonesia and Malaysia such welcoming zones for the palm oil industry, including decades of US-backed and often genocidal anti-communist counterinsurgency; the marketing tools, with their origins in the late nineteenth century soap advertisements, that convince us that, today, palm oil-derived cosmetics are "eco-friendly" and organic; the way each of us is compelled to participate in a financialized economy where the money we save or borrow passes through the same circuits as the wealth that drives the palm oil industry; the way that almost every human body on this planet has or will metabolize palm oil, transmuting it into the energy we need for work, for thought, and for action.

It is a sublime web, too complex for you or I to imagine, so it is unsurprising we create proxies, or we forget and remember, know and unknow. Yet what these charismatic proxies hide is the larger order of sacrifice looming in the background. To the extent we imagine ourselves as *merely* a consumer, an actor whose agency is limited to the scope of our market participation, we participate in the same cosmological order that demands and normalizes this sacrifice.

# Whose weapon?

Palm oil was essential to the lubrication of industrial machinery, including machines of war. The
engines and mechanisms of steamships that
allowed for the rapid expansion of the export
of palm oil to Europe were oiled with palm oil.
It greased the wheels of the railways built into
African territories to facilitate the extraction of
palm oil, minerals, and other materials. Indeed, it
was the advent of steam ships that helped cheapen
the price of palm oil in the mid-1800s, such that
it could become such a pivotal commodity, so far
from home.[105] And it was on faster moving and
more consistent steamships, lubricated by palm
oil, that seedlings of *E.guineensis* were transported
in the early twentieth century to the Dutch and
British colonies of (respectively) Indonesia and
Malaya.[106] Singapore became the shipping and
financial hub of the industry, and it remains so
today.

Part of the enthusiasm for palm oil might have
been the prospect of what might prove to be a
philosopher's stone of late-nineteenth-century
imperial warfare: a universal gun oil. Both older
muzzle-loaded rifles and newer, breach-loaded
and cartridge-firing firearms required care and
maintenance, especially in the humid tropical

climates to which empires deployed their forces, such as the Edo Kingdom.[107] Matters became more pressing with the arrival of the new industrialized weaponry of the late nineteenth century including, notably, the machine gun. The Benin Punitive Expedition was an early field test of this devastating weapon, which cut the Oba's finest soldiers—and uncounted civilians—to ribbons.[108] The heat and humidity of tropical theaters of imperial warfare meant that lubricants to prevent jamming and preserve military hardware were at a premium.

Liverpool, the British Empire's dominant palm oil port, was also home to a thriving industrial chemistry sector and importers and investors were keen to encourage experiments that might open new markets.[109] The prospect that this experimentation might bear results must have been enticing to merchants and military planners alike. Fresh in the minds of the British was the spark that ignited the Indian War of Liberation in 1857. Sepoy soldiers of the East Indian Company had been given new munitions, required they use their teeth to tear open packets of gunpowder, packets that were very likely sealed with animal fat. The prospect of Muslims being compelled to taste pig fat, or Hindus to taste cattle fat, catalyzed a long-brewing resentment among Sepoy soldiers against the Empire, leading to a revolt across the subcontinent that claimed thousands of British lives and saw many atrocities against English and Company officials and their families. However, the revenge taken by the British in the aftermath is legendary for its sadistic cruelty: hundreds of thousands were killed, and cities were razed. Torture, mutilation, and rape were policy.[110]

While palm oil may never have become the universal gun lubricant of which the Empire dreamed, it is almost certain it was a key component of an even more deadly imperial weapon. In 1865 Alfred Nobel began manufacturing dynamite in a factory near Hamburg where he could source the clay that would stabilize the synthetically produced nitroglycerine. That deadly chemical elixir was the combination of potassium nitrate (found in natural form in saltpeter or refined from bat guano) and glycerin, a by-product of the refining of fats.[111] It is highly likely that Nobel would have found the cheap, "clean" glycerin from imported West African palm oil an irresistible choice. In the later nineteenth century, Germany's import of palm oil, via Hamburg, was second only to Britain.[112]

The impacts of this invention are legendary: Nobel was haunted by the cataclysmic humanitarian consequences, spurring him to bequeath his massive fortune to the establishment of the eponymous prize to encourage peace and human achievement. In the early twentieth century, anarchists and revolutionaries in Europe and elsewhere found in dynamite a means by which the people, or a small number of their self-designated champions, might take on the great powers of empire.[113] A single, dedicated individual armed with powerful explosives could target whole squads of soldiers or police, especially in the unconventional theaters of urban warfare. Explosives could even be used to target the powerful in their homes, clubs, and retreats: "propaganda of the deed" aimed at showing the vulnerability of the powerful and breaking their fetish power over the imagination of the oppressed. The ability to efficiently sabotage rail-lines, bridges, and other logistics infrastructure appeared to place a powerful if terrible weapon in the hands of the militant oppressed. Newspapers frequently carried stories of conspiracies to use explosives, from simple grenades to elaborate bombs targeting colonial administrators, imperial monarchs, and capitalist magnates. But fear of terrorism, then as now, helped inspire and justify new forms of policing, surveillance, and repression.

But even Nobel's preferred "peaceful" industrial applications of dynamite were not without their terrors. Dynamite became crucial in blasting into the earth to directly access minerals, unleashing a hitherto unimaginable boom in mining around the world, with its attendant ecological, economic, and human violence.[114] It also enabled new methods for the levelling of land and cutting through rock to build the roads and railways to facilitate the extraction of resources from once-inaccessible interiors. Capitalist railway builders in North America and elsewhere didn't even bother to count the number of indentured migrant "coolie" laborers sacrificed to cheaply made or carelessly deployed dynamite in the creation of railways and other industrial and logistics infrastructure.[115]

Of the destructiveness of dynamite as a weapon of war, little more can be said. By the early twentieth century its toll was profound, not only as a means of killing soldiers but also targeting civilians and civil infrastructure as part of a campaign of total, industrialized warfare. When combined with the advent of military aviation that allowed bombs to be dropped from the skies, the impact was cataclysmic.

The European battlefields of the twentieth century were widely likened to grotesque scenes of human sacrifice at a hitherto unprecedented scale that demanded the lives, the sanity, and the homes of millions of working-class people. Of the First World War, Erich Fromm famously observed that, whereas in allegedly less "civilized" cultures barbarians sacrificed their own children, in the "civilized world" this barbarism was conveniently outsourced, with elder generations sanctimoniously pushing their sons into wars where they slaughtered one another. The blood-soaked trenches and haunted no-man's-lands were gory altars to a dark hybrid god of industrialism and nationalism, fueled by capitalist imperialism. "In the case of child sacrifice," he writes, "the father kills the child directly while, in the case of war, both sides have an arrangement to kill each other's children."[116]

The most famous representation of the nightmare of industrialized warfare is Pablo Picasso's *Guernica*, completed in 1937 and depicting the results of the aerial bombardment of the titular Basque city during the Spanish Civil War by German Nazi and Italian Fascist forces. The sadistic precision and destructive power of the bombing, and its targeting of civilians, would become a grim omen for the Second World

War, which was to begin two years later, when explosives killed tens of millions of people and decimated many of Europe's most populous and important cities.

Famously, Picasso had worked closely with the household and industrial paint manufacturer Ripolin to fashion a paint for his use on large canvases, and *Guernica* may have been its debut.[117] Such paints rely on resins which, by 1937, were already being synthesized with glycerin, and palm oil was a reliable and cheap source of that protean substance, though even the most sophisticated of today's methods of chemical analysis can't confirm the source of the glycerine and Ripolin's records appear to have been lost. Today, the palm oil industry and campaigners note paint as among one of the many quotidian materials in which traces of palm oil might be found. It is quite literally all around us in the built environment.

Art historically, *Guernica* represents perhaps the most famous and influential expression of Picasso's signature style. While *Guernica* might more formally be analyzed as cubist or surrealist, the trace of his primitivist turn is undeniable. Primitivism generally refers to an artistic movement Picasso helped inaugurate with his 1907 canvas *Les Demoiselles d'Avignon.* The movement took inspiration from non-Western art, especially African art, to evoke what its protagonists considered a more vivid, direct, emotive, erotic, and authentic expression. It was through the emulation and appropriation of African techniques, specifically West and Central African masks, that European painters were able to develop what was considered a revolutionary means to express the most fraught and ambivalent affects of modernity.[118]

At the core of primitivism is a kind of narcissistic racist reverence for allegedly less "civilized" cultures whose "primal" cosmology allows for a more direct and raw expression of human creative and emotional impulses, especially when compared to what many modern artists saw as the stultifying conservative norms and conventions of bourgeois European society.[119] Implicit is a progressive narrative of civilization which holds that societies evolve from "simple" and primitive origins, where people are moved by animalistic passions, tyrannical social systems, and superstitious cosmologies, toward "complex," self-reflective (if unhappy) forms of social organization, of which European modernity deemed itself the apogee. It is customary to argue that Picasso's ability to capture and predict the world-historic terror and pain of aerial bombardment in *Guernica* stems from this primitivist technique, although by 1937 it was already deeply familiar and popular, no longer quite avant-garde. Like the discourse of fetishism, the aesthetics of primitivism used a proxy version of a homogenized and defamed African culture to reveal the dangerous primitive roots of human nature which Europeans, it was said, had most successfully repressed.

There appears to be no conclusive evidence that Picasso was directly inspired by the Benin Bronzes looted by the British Punitive Expedition, but they were certainly a very prominent and celebrated part of public exhibitions and private collections of African art in the early twentieth century. Picasso attributes his inspiration to complete his *Les Demoiselles d'Avignon* to a visit to Paris's ethnographic Musée du Trocadéro's African Wing, probably in June 1907, where he was almost certain to have beheld artifacts looted from the Edo Kingdom.[120] Even at the time, within the racist European typologies of African art, the work stolen from the royal compound of the Oba was presented as at once, both uniquely "advanced" and typically savage: both singular for African art in its refined craftspersonship and yet also demonstrative of an inferior civilization and race.[121]

Perhaps because Picasso was the product and in many ways a reproducer of this Euro-supremacist milieu he vigorously denied and scoffed at questions about how African art had inspired him, preferring instead to direct attention to his own creative genius as the product, indeed the climax, of European modern aesthetics. It was a story contemporary critics were all too happy to echo. Here, a certain notion of art served to reinforce a global white supremacist cosmology where the figure of the artist mirrored and flattered the figure of the entrepreneur, *homo economicus*: the self-willed, individualist actor who forges order out of chaos.[122] Picasso was a communist and, in his way, supported some anti-colonial struggles. Yet the persona he adopted and promoted was of a piece with the cosmology of the market.

Such a cosmology envisions itself as the triumph of the world-historical event of European modernity, the only society to allegedly have transcended superstition through the application of self-reflexive reason. But this cosmology permitted and relied on a world of heinous sacrifices.

In 1942 the Americans developed a weapon that would become the icon of neo-imperialist warfare in the latter half of the twentieth century by combining petrochemicals with palmitate, an acid derived from several natural fats but, as the name suggests, primarily from palm oil: napalm.[123] While napalm was widely used by the US in the Second World War, its appeal became most famously apparent in the US empire's campaigns of counterinsurgency and asymmetrical warfare, such as the American invasion of Vietnam, though by that time other chemicals provided the gelatinous base that clung to the skin of victims, a living nightmare. It was the profound success of anti-colonial struggles around the world that encouraged American and other militaries to use this vicious weapon. While they were usually no match for imperial troops on the conventional battlefield, these struggles could win significant victories through guerilla warfare, turning to tropical forests and sympathetic locals as strategic assets. By valuing local people and land as allies, insurgents sought to make war so expensive to the occupier that they quit the territory, opening the possibility for true self-rule in the wake of decades or centuries of imperial domination or neocolonial wealth extraction. By contrast, napalm is famously inaccurate and was indiscriminately dropped from helicopters and planes as part of a scorched earth strategy. Unlike a machine gun or explosives, it is difficult for guerilla insurgent forces to appropriate or redeploy, as they typically lack air forces and

# Whose fat?

You and I are likely most familiar with palm oil as a food and, these days, as a food to be avoided, either for ethical reasons related to the horrific conditions of its production or for reasons of health. This includes its high concentration of saturated fat, which is widely recognized to have adverse effects on heart health and can contribute to diabetes.

Palm oil was not widely consumed for nutritional purposes outside of Africa until the twentieth century, in spite of entrepreneurs eager to transform it into a cheap and plentiful fat to feed the nineteenth century's swelling urban proletariat for whom staples like butter and lard were increasingly too dear to afford.[124] The commodity's common association with locomotive and axle grease made it a tough sell, though by the turn of the twentieth century it was being used to feed cattle and adulterate butter exported widely from the Netherlands and Germany.[125]

But palm oil was pivotal to the manufacture of another nineteenth century imperial commodity: canned food. The canning of food was initially a military and imperial experiment. The prospect of being able to, in a safe, durable, and efficient way, preserve meat for long naval voyages was tantalizing to admiralties keen on global conquest.[126] Until the mid-twentieth century, palm oil was the essential ingredient in the manufacture of cans, providing the flux in which thin sheets of rust-proof tin were fused to iron before being shaped into cylinders for packing.[127]

From the middle of the eighteenth century until well after the Second World War, English army and navy rations were predominantly canned. In part, this was prompted by a desire to ensure the enforcers of empire would not need to depend on local food in their zones of deployment, for fear of both ruthless profiteering by middle-men and sabotage from anti-imperialist locals. A mass poisoning of European civil and military personnel occurred in Hong Kong in 1857, an act that saw the survivors and their backers in London rattle the sabers of vengeance, although they conveniently forgot the context: The British Empire was using Hong Kong as a base of oper-ations for the opium trade that was poisoning a shocking percentage of the Chinese population and virtually enslaving hundreds of thousands of Indian peasants and workers to grow and refine the poppies that fed the insatiable and deadly trade.[128] London's Royal Society conducted stren-uous research into poisons and their antidotes in the service of empire.[129] With so many poten-tial enemies working in kitchens throughout the empire, canned food was key to the logistics of maintaining empire. The *Malacca*, the naval ship that carried British marines and troops to Benin for the 1897 Punitive Expedition, for instance, was outfitted for the voyage with 2,200 tins of canned milk, 200 tins of Bovril and 120 tins of meat extract.[130] Those cans almost certainly emerged from a palm oil bath.

Like soap, canned food became a key means by which notions of empire were broadcast and the poorer classes seduced into fidelity to empire. While at first canned goods were marketed to the aspiring consumers of the middle class as a cosmopolitan luxury, canned food's ability to provide cheap nutrition soon made it a mainstay in working class nutrition. When canned foods were first marketed to the public, they highlighted the products of empire: salmon from Canada, meat from Australia, and exotic fruit from the tropics. Consumption was promoted as a patriotic duty. As Geographer Simon Naylor demonstrates, the modern miracle of safe, durable, and globally sourced canned food was presented to the public as the triumphant imperial conquest of space, time, and nature.[131] But at the same time, the prominence of canned food and its close association with empire made it a prime target for boycotts.[132]

It was only in the later nineteenth century, after French chemists discovered how to hydrogenate vegetable fats, that Europeans began consuming palm oil directly as an ingredient in margarine. Such consumption accelerated in the wake of the militant strikes of Chicago meat industry workers around the turn of the century, which drove up the price of animal tallow.[133] The market for vegetable margarine increased by a factor of four from 1875 to 1900 and apace during a half-century of war, depression, and war again.[134] Palm oil at various times and in various places was a key ingredient in this foodstuff.

Today, bleached, deodorized, and highly refined palm oil is eaten in one way or another by hundreds of millions of people around the world every day. The widely cited estimate that palm oil or its derivatives are present in over 50% of supermarket products reveals that its ubiquity goes well beyond the huge variety of snack foods and processed edible products that use it as a cheap, tempting, and durable fat: its trace or byproducts are in a vast range of foods and can appear under some 200 different names. In the late twentieth century, both dairy and soy industries waged a campaign to paint palm oil as a nefarious, toxic, and shadowy tropical invader in the American diet and to promote partially hydrogenated products from other plants as replacements.[135] But the discovery of the disastrous health impacts of trans fats in these products in the 2010s saw a swing back to palm oils, which, like butter, retains a semisolid consistency at "room temperature."

For the sake of simplicity, I am using the term palm oil to encompass all products that derive from the fruit of the oil palm, whether from the fleshy bulbs or the hard inner kernels. Palm oil today is not the most commonly consumed fat globally, but it does fill a very particular set of niches. In contrast to the saffron-hued and pungent character of West African virgin or red palm oil, RBD (refined, bleached, and deodorized) palm oil and many other industrially produced products are neutral, colorless, and largely tasteless, making them a tempting ingredient or additive to a vast own nutritional reproduction. For a planet where half of all human beings live or will soon live in "slums," (poor, informalized urban or semi-urban zones), foods like palm oil are crucial.[137]

In India, for instance, as traditional lifeways are evaporating in the face of corporate-led modernization, consumption of processed packaged foods has increased by almost 20% every year since 2014, with dramatic impacts on poor people's health.[138] Palm oil has slowly become the primary cooking oil for poor people in not only India but around the world. In Indonesia it represented 2% of cooking oil in 1965, but 94% in 2010, thanks in no small part to the government-supported marketing campaigns that framed it as not only a nationalist economic miracle but an intimate part of daily life.[139] The cheapness of palm oil is also seeing it replace local, artisanally produced oils, further contributing to poor people's reliance on this globally traded commodity. By-products of the palm oil refining process are also increasingly sought after as additives to animal feed for both domestic pets, farmed fish, and livestock.[140]

For a vegetal product, palm oil is strikingly high in saturated fats and its health impacts are strenuously debated, with eager participation by the industry itself that has everything to gain from promoting it as a nutritious and safe food.[141] But regardless of biochemical questions, the sheer abundance of cheap cooking fat that palm oil offers, as well as the relative cheapness and convenience of oil-fried packaged foods, have all contributed to profound increases in heart disease, diabetes and other chronic health ailments in regions where it is widely consumed. In India, the world's largest palm oil importer, where it is typically consumed by the poorest people, heart disease is threatening to reach epidemic proportions.[142] One study found that, in developing nations, every additional kilogram of palm oil consumed per capita was correlated to an additional 68 heart disease deaths per 100,000 people.[143] Recent revelations about palm oil's carcinogenic qualities have also caused great concern among wealthy consumers and epidemiologists, but poor people have few other options and often lack access to relevant information.[144]

Palm oil's cheapness has little to do with the oil's natural properties and everything to do with the social, cultural, political, and economic circumstances of its production. *E. guineensis* is, by reliable though not undisputed calculations, the most productive oil-bearing plant per acre when the costs of all inputs are factored in: Today, palm plants generate 30% of the world's edible oil but occupy only 5% of oil-producing land.[145] But much of this calculus depends on taking for granted the profound cheapening of both land and labor in palm oil growing zones, as well as the ability of plantations and smallholders to externalize costs, for instance by ignoring the ecological impacts of rainforest destruction or the toxic runoff of agricultural chemicals.[146] It is cheap, too, because the labor to grow it is cheapened. Unlike canola or soy oil, which derive from relatively small plants, close to the ground, the cultivation of palm fruit, which grows high up in thorny trees, is almost impossible to automate, requiring considerable human labor. Hence its cultivation as a cheap food is only possible if the labor is made cheap.

Today, Malaysia and Indonesia are host to the vast majority of *E. guineensis* and the infrastructure to transform its bounty into many useful things. The plant arrived there in the mid nineteenth century on European ships. It first served as a stately ornament, lining the pathways that led to the mansions where Europeans presided over migrant laborers toiling to clear forest to grow rubber, sugar, and other cash crops to feed the imperial furnace. Here, the strategies of the plantation, elemental of capitalism since its inception, were applied at industrial scale.[147] These included the seizure of Indigenous and peasant lands, clearcutting of forest, the exploitation of dispossessed laborers, divide-and-conquer techniques to pit workers against one another, the use of debt bondage and company store tactics, and the mobilization of gangs and paramilitary forces or "private security" to repress resistance.[148] To these were added innovations in the management of joint stock corporations and transnational financial markets, which facilitated cooperation between metropolitan investors and frontier entrepreneurs. Then as now, the corporate structure perfects a kind of moral and political separation, creating an unscrupulous artificial entity oriented exclusively to returns for investors, allowing those who fund and profit from such operations to wash their hands of these horrors, claiming they are the will of the omnipotent market.[149]

In spite of the magnitude of violence unleashed, resistance was profound, even if it was not always recorded by the European planter class or their local elite clients.[150] Strikes and even open revolt plagued the plantation economy in South East Asia. The dense forest provided strategic cover for those revolting migrant workers whose energies would otherwise be dedicated to its destruction.[151] Like any and all resistance, theirs was neither purely heroic nor purely villainous: some sought modest improvements in working conditions, others land rights. Later, in both Malaysia and Indonesia, powerful anti-colonial struggles drew on generations of militant labor struggles to wage devastating guerilla warfare against British, Dutch and Japanese occupations. After the Second World War, many such movements took the form of communist insurgencies that were monstrously repressed by newly independent states with help from imperialist powers.

It wasn't until the twentieth century that South East Asian plantations transitioned toward palm oil, a crop that thrived in environments, notably peatlands, that were inhospitable to more profitable crops. In 1934, Indonesia, still a Dutch possession, outpaced Nigeria as the world's largest exporter. It was overtaken by Malaysia in 1966, thanks to an aggressive set of nationalist government incentives that sought collaboration with European plantation companies and buyers.[152] Malaysia's boom was abetted by growing post-war markets for processed snack foods, and then again in the 1970s by rising petroleum prices that encouraged chemical companies to turn to palm oil as a cheaper alternative.[153] Careful collaboration between the Malaysian government, national and regional elites, Western corporations, and the World Bank, saw palm oil plantations grow and exports increased by a factor of 25 in the two decades leading to 1984. This was partly also thanks to the establishment of onshore palm oil refineries and oleochemical plants.[154] Similar factors allowed Indonesia to regain its status as the world's top exporter in 2006, an industry stewarded under the vicious authoritarian regime of Suharto and which spread in the crony-capitalist and kleptocratic climate left in the dictator's wake.[155] It is within these three decades that a new oligarchy of profoundly powerful South East Asian corporations arose that, today still rule the global palm oil industry, alongside longtime colonial-era interests like Unilever, Cargill, and SocFin.[156]

Despite the relatively few players that refine and export palm oil, the global market for palm oil is highly diversified: the palm oil purchases of the 137 leading retail, food-service, and manufacturing companies only amounted to 10% of global production.[157] The other 90% represent a vast array of intermediaries and end users. This is among the reasons why, even though it is traded as a commodity on international financial markets, palm oil is fairly marginal to and insulated from the vertiginous speculation that has plagued many other basic foodstuffs in the past decades.

While it represents a relatively minor player in global palm oil production compared to South East Asia, the Latin America industry has witnessed similar forms of injustice relative to its industry, though with different expressions in different nations, from Peru to Honduras, from Brazil to Guatemala. While there was a short-lived attempt to develop palm plantations in Brazil in the early twentieth century, most of the continent's experience with industrial cultivation of palm oil has come in the last two decades, and often in the wake of profound violence, much of it connected to the operations of the hegemonic America United Fruit Company.[158] Throughout Latin America, palm oil has appealed to landholders who are seeking to diversify their operations and who can benefit from cheapened migrant labor forces, many of them displaced by decades of conflict or destructive mining and extractive projects. In the Amazon and elsewhere, entrepreneurs have cut deeply into forests or seized Indigenous and peasant lands to establish new plantations.[159] More widely, palm cultivation occurs on already cleared lands that may have previously grown other crops, but often in contexts where decades of political violence places landowners and corporations at an often deadly advantage relative to workers, and where environmental laws are lax or poorly enforced.[160]

It is difficult to overstate the palm industry's profound destruction and transformation of human and non-human life. Anthropologists Tania Murray Li and Pujo Semedi liken the corporate plantation system in place in Indonesia to a military occupation and a mafia operation, with devastating impacts on human rights, economic justice, and ecological vitality.[161] In South East Asia's "corporate food regime," Pye sees an extreme version of the kind of exploitative proletarianization that has long marked capitalism's global development.[162]

A handful of prominent recent reports sketch the contours: A 2016 Amnesty International study of the Indonesian operations of Singapore's Wilmar International, the world's largest processor of palm oils (controlling 43% of global trade), found "serious human rights abuses on the plantations" of the company and its subsidiaries (of which there are over 300) and suppliers.[163] These included "forced labour and child labour, gender discrimination, as well as exploitative and dangerous working practices that put the health of workers at risk." Further, these abuses "were not isolated incidents but due to systemic business practices . . . [including] the low level of wages, the use of targets and 'piece rates' (where workers are paid based on tasks completed rather than hours worked), and the use of a complex system of financial and other penalties," that unduly affected casual and migrant workers. A 2013 study by Indonesia's Sawit Watch, which researches and advocates on behalf of palm-oil affected communities, and the Washington-based International Labor Rights Forum of three sites in Malaysia and Indonesia "found flagrant disregard for human rights at some of the very plantations" that the industry's voluntary watchdog, the Roundtable on Sustainable Palm Oil (RSPO), classifies as "sustainable." Abuses included "labor trafficking, child labor, unprotected work with hazardous chemicals, and long-term abuse of temporary contracts."[164] At the end of 2020, and after months of investigation, the United States Customs and of their land in name only and have been made reliant on financing from corporate actors higher up the production chain, or from local landlords and moneylenders, relations that conveniently outsource oppression and exploitation and providing the veneer of entrepreneurial opportunity.[168] Certain aspects of the industry have been heavily mechanized, but this has typically had the effect of again concentrating power in the hands of those who can afford machinery. It remains the case that industrialized palm oil production relies on a large amount of deskilled human labor, and often whole families are conscripted to cultivation, harvest, and transport of the product, including children, although often only the male head of the household is paid. The gendered dimension of exploitation is compounded by frequent incidents of sexual assault.[169] Meanwhile, workers are routinely exposed to harsh and often toxic chemicals that are used in industry and are also often tasked with doing highly dangerous work, including cutting fruit bunches from high up in the canopy or otherwise working in hazardous environments.[170] In both South East Asia, Latin America, and West Africa, the Cold War cast a long shadow: Often the very zones that are abundant in palm oil cultivation are where, a generation or two ago, CIA-trained or -funded death squads carried out their brutal harvest. In many of these places, the forms of lawless violence, corruption, and human and environmental disposability continue to plague the population.

Part of the problem is the relative ease with which the palm oil industry can expand, and the occurs in isolation, often operating alongside intensive forestry, mining, and infrastructure development. But in South East Asia, almost all studies agree it has been catastrophic: plant diversity on our near plantations has in some cases decreased by 99% and mammal diversity by 47–90%, with similar impacts on other forms of life.[174] Some estimates suggest palm oil plantation expansion adversely affects 54% of threatened mammals and 64% of threatened birds globally.[175] Outside of West Africa (which represents a tiny fraction of global production), *E. guineensis* can be considered an invasive species, and it has had dramatic consequences on local ecosystems. Palm oil plantations also contribute to erosion and often leech pesticides and herbicides into the water table. Oil palm cultivation has so drastically transformed the landscape it has affected the weather in South East Asia, changing rainfall patterns.[176]

For many Indigenous groups, the destruction of the rainforest has eliminated food autonomy, traditional medicines, and endangered or erased whole ways of life. Workers in the industry also suffer from the environmental destruction to which they are compelled to contribute. One demonstrative study revealed that palm oil workers tend to eat far less diverse and healthy diets than others of a similar background in the area, diets that are ironically high in industrially manufactured foods, high in palm oil, notably instant noodles.[177] As Pye argues, the palm oil industry appears as a giant machine for producing a new precarious proletariat, alienated from the land on which they live and toil and, ironically and tragically dependent on refined palm oil and other industrially produced foods for their survival.[178]

Almost all of these humanitarian and environmental dimensions took a turn for the worse in the last 15 years thanks to burgeoning markets for biofuels. Then-US president George W. Bush's Energy Independence and Security Act of 2007 and the European Union's Renewable Energy Directive of two years later set targets for the proportion of fuels to be derived from "renewable" sources. This prompted the governments of palm oil exporting countries to promote the expansion of the industry to meet the rising demand for agrofuels.[179] Major international financial institutions and longstanding palm oil players rushed to not only generate more supply but also promote palm oil as a sustainable solution to the climate crisis. The result was a massive expansion of the industry that encouraged dramatic deforestation.[180]

The horrors of the palm oil industry have, of course, not gone unnoticed. Understandably, over the past three decades, major environmental NGOs have targeted the palm oil industry for special attention, including large organizations like Greenpeace, the Rainforest Action Network, the World Wildlife Fund and Amnesty International. But this focus on palm oil is at least partially due to the charisma of such campaigns in the Global North. Since at least trace amounts of palm oil are in a huge variety of products used by consumers on a daily basis it offers environmental campaigners, many of whom work for organizations that depend on donations, a tempting point of contact with the general public. For NGO campaigners and the media, the image of Western corporations, many of whom emerged from the cradle of colonialism, using "conflict palm oil" produced under disastrous conditions is appealing in the fast moving attention economy. Unilever, for example, which remains one of the world's preeminent manufacturers of both packaged foods and cosmetics, has been a favorite target, and for good reason. Perhaps the most famous campaign was waged by Greenpeace in 2008 and subverted the brand's own marketing tropes that highlighted the firm's allegedly ethical stance on body image. The campaign included publicity stunts with protesters dressed as orangutans, all calculated to draw attention on social media.[181]

Yet this attention has also seen Unilever and other brands work with the exporting industry to rebrand their sources of palm oil as "sustainable" and to lionize "socially responsible" corporations as champions of human rights. In order to combat bad publicity from its competitors and non-governmental organizations, the palm oil industry has secured the services of many of the world's most notorious public relations firms, including those infamous for their service to the petroleum industry and cigarette companies and who deploy remarkably similar tactics: casting doubt on science, promoting their clients as reformed corporate citizens, or cynically focusing public attention on companies' small, charismatic social, or environmental projects to the exclusion of the larger, more damning picture.[182]

The industry has been adept at pivoting its rhetoric to seek to capture the dominant global ecological discourses without significantly changing its activities. This has included many climate schemes that reframe the expansion of the palm oil industry as of net benefit given that, from a very limited perspective, oil palms, as plants, draw carbon from the atmosphere.[183]

The Roundtable on Sustainable Palm Oil (RSPO), established in 2004 in response to mounting global civil society pressure, is a global organization intended to both regulate and promote the palm oil industry. In addition to industry and government representatives, it also includes some of the largest and most influential environmental NGOs, notably the powerful World Wildlife Fund, whose influence in France and Germany, where palm oil consumer activism is perhaps strongest, in part thanks to their efforts.[184] Scholars, local activists and smaller NGOS have greeted this cooperation with deep skepticism and even anger.[185] Indeed, in spite of what must be acknowledged as significant improvements to mainstream palm oil cultivation and refinement practices, the RSPO has been largely ineffective at actually stopping deforestation, the seizure of Indigenous land, the loss of species habitat and biodiversity, or the abuse of migrant workers. In part this is because so many of these occur in the murky territories near the origins of the pro-duction chain, often in the hinterland frontier where sub-contractors and sub-sub-contractors operate with near impunity and no oversight.[186] Yet abuses continue even within the estimated 21% of all palm oil operations certified by the RSPO as "sustainable," thanks to a combination of lax enforcement, sluggish auditing, and the scale and complexity of the industry's operations.[187] The broader context is the ever increasing demand for palm oil in global markets, facilitated and promoted by those larger corporations and companies who profit from infinitely expanding markets, demanding ever more land be cleared and ever more cheap labor.

Targeting the palm oil industry, then, offers NGOs in the Global North a convenient foreign villain who, ultimately, has little power and influence in the nations where those NGOs are headquartered. This has not been lost on governments and representatives of palm oil businesses in palm oil producing countries, who, largely cynically, accuse these NGOs of sanctimonious neo-imperialism.[188] The recent high-profile (and not illegitimate) concerns raised by Western celebrities about the impacts of palm oil, or the high-minded campaigns of Western grocery stores or food and cosmetic brands to ban palm oil have led to a popular nationalist backlash in Malaysia and Indonesia, whipped up by the industry.[189] Unfortunately, both the focus on the actions of international companies, governments, and NGOs has also been used to diminish or sideline the work of local organizations in palm oil producing countries. Many of these are congregations of workers, of migrants, of Indigenous people, or of smallholder farmers who have come together to contest the humanitarian and ecological conditions of the industry. Some have been remarkably effective not only in their limited locality but in the wider region as well, thanks to solidarity efforts.[190]

Yet these organizations and individuals are often made to pay the price for the backlash triggered by more high-profile campaigns, often with their lives. The murder of journalists, trade unionists, and human rights and environmental campaigners is unfortunately nothing new in the palm oil industry.[191]

But these contradictions have not stopped spirited resistance to the industry.[192] In 2021 alone, multiple Indigenous groups around the world, including in Guatemala, in Indonesia's Papua province, and in the Brazilian Amazon, have used a variety of methods to defend their ancestral lands from land grabbing, ranging from partnering with global NGOs to wage court cases and public relations campaigns overseas to non-violent disobedience.[193] In Malaysia, Indonesia, the Philippines, and Columbia, labor activism is fertile and increasingly coordinated, though organizers are often subject to abuse, threats, and even murder.[194] In West and Central Africa, local communities have been protesting land grabbing to make way for large plantation estates and, instead, valorizing traditional artisanal production.[195]

The story of palm oil is difficult to tell because responsibility for its horrors is slippery. Who should be held accountable for environmental destruction, the abuse of migrant workers, the theft of Indigenous lands, the murder of activists or the annihilation of whole species? Local landowners and plantation managers point to pressures from above while corporations plead ignorance or helplessness for what happens deep in the jungle. Governments claim to be trying to encourage best practices without endangering a major source of economic growth and international competitiveness. Transnational buyers, including prestigious and brand-sensitive corporations, point the finger at their suppliers and note (with some degree of honesty) the near-impossibility of monitoring their entire supply chain while still providing affordable products to consumers and satisfying shareholders. Within this tangle, much attention has focused on the smallholder.

On the one hand, the palm oil industry and its allies in environmental NGOS, along with governments and the media, have been keen to celebrate the smallholder farmer as the beneficiary of the booming market and the potential victim of overzealous regulation or oversight. Certainly there is ample evidence that smallholders can efficiently and reliably produce impressive quantities of oil without most of the heinous environmental and labor atrocities that otherwise characterize the industry, and can do so in ways that preserve workers' autonomy, dignity, and a reciprocal relationship with the land.[196] But the definition of smallholder is rarely exact, and often easily gamed.[197] Ironically and conversely, the fetishization of the smallholder—said to be responsible for 40% of Indonesia's oil palm cultivation—has allowed industry advocates to turn the narrative around, arguing that while large refiners and exporters of palm oil would *like* to be able to enforce labor and environmental regulations, they cannot constantly be auditing wily smallholders who do as they please in the rainforest.[198] A similar double movement can be observed with the way these forces frame Indigenous people, sometimes as victims in need of protection, sometimes as greedy parochial primitives incapable of following global rules for their own good.[199]

Thus, it has long been in the palm oil business, which is in some ways a revealing microcosm of the role of entrepreneurialism and investment in capitalism as a whole. Until the twentieth century, European traders were, by and large, content to trade palm oil in West Africa from the safety of their ships or coastal factories, leaving the business of extracting and transporting the crude palm oil to African traders.[200] This system was both enabled and enforced by the European use of "trust" schemes of credit that advanced their African partners' trade goods and weapons in return for future payback.[201] Debt, then as now, was used, intentionally and unintentionally, as a tool of extraction and control. Ultimately, this system benefitted the merchant capitalists and financiers back in London and Liverpool, who were able to wash their hands of the business of slavery and hyper exploitation located well down the financial chain.

Though the political economy of palm oil is in many ways unique, it perhaps reveals something more universal: capitalism as a global system compels and encourages every actor to adopt a competitive disposition in the name of survival, from independent farmers to migrant workers to factory bosses to corporate executives to financial traders to company purchasers to marketers to store managers to consumers and even to NGOs. If people and the planet are sacrificed on the altar of the industry it is because it is in many ways emblematic and constitutive of a broader form of sacrificial capitalism that makes almost everyone a little bit complicit, though some more complicit than others, and with very different costs and consequences. The palm oil story reveals a global web of coercion, cooperation, and competition in which we can see the outlines of the interconnections of a now-global species. The nature of these interconnections, at the macro level of the system as a whole and the micro level of the motivation of individuals, is shaped by the "fetish power" of the market, which we take for normal, natural, inevitable, and unquestionable. Yet hiding behind it, in plain sight, is the raw elemental cooperative potential which might, ideally and in the future, be orchestrated otherwise, toward more peaceful and abundant horizons.

# Whose surplus?

In order to justify the constant expansion of the palm oil industry, its proponents often play on common anxieties of a growing global population in need of affordable nutrition.[202] These discourses implicitly or explicitly cite myths of overpopulation that have their origins in racist eugenics and are frequently used to justify or rationalize the sacrifice of people—typically racialized or otherwise stigmatized people—in the name of the abstract greater good.[203] What such discourses render invisible in the broader capitalist socioeconomic contexts in which populations and resources become abundant and scarce, as well as the vast differentials in consumption between the world's rich and poor.[204]

Within this context, palm oil feeds hundreds of millions of people who are cast to the bottom of global hierarchies of income and security and who are often excluded from the formal capitalist economy completely, such that they can only afford this artificially cheapened fat. Their own socially-embedded physical reproduction, the metabolism of food into energy to keep a heart beating or a mind thinking, depends on palm oil. The *E. guineensis* metabolism depends on the conversion of sunlight, water and soil-bound nutrients into fleshy seedpods. Capitalism metabolizes these seedpods into cheapened food than can be consumed a world away. It is a chain of metabolic processes which begin in sunlight and ends in human labor power, that is mediated by capitalist relations and the artificial scarcity of money that compels each individual along the chain to make choices they otherwise might not make or might make otherwise.

For Georges Bataille, all earthly societies practice sacrifice, based on the universal fact of the sublime, near infinite magnitude of the solar radiation responsible for almost all life on earth. Whereas a long lineage of European economic thinkers presumed that all society is necessarily about the management of scarcity, in Bataille's heretical view the problem was, rather an encrypted abundance: There is too much energy, too much life, so much that it jeopardizes the legitimacy of systems and structures of power based on the presumption of scarcity. Here, wealth and scarcity appear as human constructs, terms defined not by objective material facts but that emerge from the way a society is organized. Objects and people are made valuable and defined as scarce in ways that justify and reproduce social power. This arbitrary social order of wealth and scarcity is threatened by the undeniable reality of solar abundance and so societies necessarily invent rituals and practices to relieve themselves of wealth through sacrifice. Yes, of course: wealth materialized in precious objects or embodied in valuable animals, whose sacrifice was said to please the gods. But also, the sacrifice of that most invaluable thing, the creator of value itself: people. For Bataille, a witness to two world wars and much else besides, human sacrifice was far from abolished in European modernity, it had only taken new forms. In a society convinced of the inevitable fact of scarcity, sacrifice became a public secret, exercised and justified through the cold logic of the market or the pious justice of the state.

Perhaps in all societies human sacrifice dare not speak its name. Perhaps sacrifice always cloaks itself as a voluntary offering, a cosmic inevitability, or a just punishment. Under capitalism, sacrifice arrives wearing the crown of hyperrational, unforgiving market necessity. Such an approach would help us explain the cruel irony that defines our world: as a species, by most material measures, we are wealthier than we have ever been. And yet, millions suffer from malnutrition and other maladies due to a shortage of reliable food, and hundreds of millions more must make do with the barest of essentials.[205]

Palm oil, it might be argued, is the fat of the world's poor. Global palm oil production has increased by a factor of 12 between 1981 and 2015. Some 70% of that ends up in food products. The industry estimates that fats will represent some 45% of all additional global calories consumed globally between 2011 and 2030, with palm oil being the cheapest, most versatile, and most reliable by a considerable margin.[206] True: palm oil can still be found in a wide variety of upmarket products, especially cosmetics, where dubious certification schemes allow it to be labelled by its derivatives and so never directly named. It can also often be classified as "organic" to appeal to wealthier consumers. Indeed, while palm oil has, as we have seen, been a major source of soaps and cosmetics since the nineteenth century, more recently it has benefitted from a turn toward "ethical consumerism" that prides itself on "vegan" products and certifications. But this does not change the fact that the recent stigma attached to palm oil, which largely stems from news of the disastrous human and environmental costs of its production in Indonesia and Malaysia, has only consolidated the notion in the Global North that those who can afford to do so should opt to avoid it. This impulse was reflected in the French government's aborted "Nutella Tax" levied on palm oil product imports, in part to put pressure on exporters to improve environmental conditions, in part to discourage consumption of unhealthy processed food products that tend to be consumed by poorer people.[207]

What such measures assiduously avoid is the complex relationship between fat and poverty: far from the familiar nineteenth century association of corpulence with the overindulgences of the wealthy, today fat is associated with the "undeserving" poor who are framed as being unable to adopt the appropriate degree of neoliberal self-control.[208] Prohibitions and taxes on edible fats, like those on sugar, play on cultural anxieties that associate fat with feeble morality and lack of self-control. Such myths hide the ways that those actually enduring poverty are systemically denied the disposable income, the time, or the other resources necessary to pursue more socially acceptable diets or other healthful activities.

Fat is an essential and important part of many organisms, especially mammals like humans who have evolved to use considerable quantities of fat to regulate body heat and store energy. But in the cultural climate of late capitalism, fat is typically framed as a kind of useless and dangerous surplus, a misbegotten and unhealthy growth. As scholars critical of the pseudoscience of "body mass index" and other phenotypical measurements of fat indicate, the appearance of what is deemed excess fat on the body is highly stigmatized in ways that clearly go well beyond an often sanctimoniously proclaimed concern for an individual's welfare or even public health.[209] This false concern is even more galling when it fails to look to the socioeconomic conditions of fat that see it accrue disproportionately to those who live in poverty or precarity. The stigma around fat is in many ways the product of aesthetic norms and forms of bodily measurement forged in European racial science in the 19th century that carry such norms into the 21st century.[210]

At stake in the depiction of both fat and the world's poor is a pathologization and fear of surplus flesh that at once strips the bearer of that flesh of value and makes them responsible for managing risks that the immiserating system has generated. Fears of "overpopulation" and of an "obesity epidemic" both profoundly but profitably misunderstand the underlying problem, rendering invisible the form of global racial capitalism within which both human reproduction and individual metabolisms transpire. In both cases, the concern is not that there is too much flesh, but that the flesh is of the "wrong" type.

The norms and beliefs around body size and shape and the ideology that animates concerns about overpopulation are historically defined, which is to say they transpire within the fields of power. Today both occur within the context of a form of global capitalism in which the basic logic of accumulation has supplanted or is profoundly challenging any and all other forms of social organization and economic value.[211] Here, the value of a body or a population is largely defined by its ability to facilitate the generation of profit. The career of palm oil helps us connect the fate of both individual bodies and whole groups of people.

Arguably, palm oil has powered the world's greatest ever migration.[212] From the 1990s until the present, no fewer than 200 million people have moved from China's small cities and villages to industrial cities, largely in the country's Southern provinces, in order to seek work. Thanks to a set of regulatory relaxations, incentives, and government programs inaugurated beginning in the late 1970s under the premiership of Deng Xiapoing, China has become the workshop of the world, with whole cities emerging to accommodate new factories and dormitories.[213] This monumental population shift, which was officially frowned on but unofficially encouraged by the Chinese state, saw many fortune-seeking migrant workers encounter extreme exploitation and privation, including conditions of debt bondage, chronic overwork, and prohibitions against forming independent unions.[214]

Within this context, instant ramen-style noodles became a vital commodity. Known now around the world, these single-serve packages of dehydrated noodles and synthetic flavors are famous for their almost infinite shelf-life, cross-cultural popularity and extremely cheap price, offering a reliable 375 calories of refined white wheat flour and vegetable oil, typically palm oil, delivered with a hit of sodium and mono-sodium glutamate.[215] By some estimates, over 100-billion servings of instant ramen noodles were consumed globally in 2017, perhaps more than any other single packaged product.[216]

These prepackaged instant noodles have their origins in the cheap price for wheat and fat (in the original case, soybean oil) dumped by the US on its East Asian protectorates, South Korea and Japan, as American aid after the Second World War and the Korean war.[217] Since then, prepackaged instant noodles have become a staple of migrant workforces around the world. In China, the prepackaged instant noodle has become a fixture of life for the country's poorest people and especially in the dormitories and cramped apartments of migrant workers. In this sense, it powered and continues to power the production of a huge proportion of the world's material culture that flows outward from those factories. Nearly everything that surrounds me as I write to you emerged from or contains components built in coastal Chinese plants where palm oil-heavy instant noodles form a fixture of young migrant workers' diets. Dozens of parts of the computer on which I type, the wax on the table on which it sits, the small plastic bottle of hand cream, the shoes I am wearing and so much more.

Though it represents only a fraction of global instant noodle consumption, my interest is drawn to often sensationalized media stories about packets of "soup," as they are known, becoming the "new currency" in American prisons.[218] Many of these stories focus on the quirky dimensions of this phenomenon, with an eerie echo of the European fascination with the fetishistic trade practices of African and other non-European peoples that seemed to incorrectly or idiosyncratically value certain "worthless" goods over truly valuable ones, notable because of the disproportionate incarceration of people of African descent incarcerated in those prisons.[219]

In these institutions, many of which are either run by corporations or rely on corporate services, the costs of incarceration are increasingly passed on to incarcerated people and their families.[220] Prison diets, which often come from surplus from the US military or the materials deemed unsuitable for sale by their manufacturers, are frequently inedible or insufficient. In a typical prison "bagged" meal (bread, processed cheese, processed meat, a cookie, drink mix and condiments), it is almost certain that every single component contains palm oil. Many imprisoned people therefore rely on supplemental food to ensure they can replenish their bodies, especially in a context where physical labor, paid at dramatically depressed prison wages, is the only way to gain any measure of financial independence and where many incarcerated people put a priority on building their strength through weight training, either as a means to maintain their physical and mental health or, horrifically, so that they can protect themselves from guards and other prisoners in a system characterized by routine structural and non-structural violence.

In contrast to inconsistent and often inedible prison rations, instant noodles, which are sold through typically privately run prison commissary offices at a massive markup, offer reliable and safe access to calories. This, along with their near-infinite shelf-life and pre-packaged uniformity, have made them a popular means of exchange and store of value in prison life. Anthropologist Michael Gibson-Light details the ingenious ways in which incarcerated people create parallel economies based on soup, including complex networks of debt and mutual aid. He also illustrates that this economy is a form of resistance to an increasingly punitive neoliberal model of mass incarceration. In the past the politics of food in prisons was a site of contention "through demonstrations like chow line boycotts, sit-ins, or hunger strikes [… but] such overt tactics have grown relatively infrequent in the contemporary penal institution." Instead, "today's prisons are home to differing degrees of strategic resistance, with a growing emphasis on covert practices aimed at diluting or circumventing prison power structures." He notes that "subtler expressions of power and autonomy— such as secreting, hoarding, or preparing food against institutional regulations—have grown more frequent and effective."[221]

For Gibson-Light, this shift in tactics of resistance is intimately connected to the changing nature of American penal structures toward mass incarceration. "Under the 'old penology' of the twentieth century, penal policy and administration regarded offenders as individual wards of the state, emphasizing diagnosis and treatment," which, he suggests, helped give rise to the popularity of the cigarette as a trade commodity with its connotations of "a withdrawn demeanor" and "assertion of autonomy in the face of treatment and responsibility."[222] The cigarette represents a medium of refusal to a biopolitical regime that justifies itself in the name of "making life" and reforming the behavior of the prisoner so as to refashion them as a functional member of a capitalist society, a self-responsible worker. But "with the development of a new, neoliberal penology . . . prisoners have come to be treated as aggregate groups and consumers of services." Hence resistance has shifted to the field of nutrition. To consume and transact with soup, which is reliably if insufficiently nutritious, is a form of refusal against a sacrificial regime. As American prisons are increasingly seen as institutions not for the reform of criminalized individuals but essentially for the warehousing of human beings and the privatized extraction of wealth from their misery, resistance often shifts to the level of life itself, or, perhaps more accurately, the field of slow death.

In Ruth Gilmore Wilson's landmark theorization of the prison as a vehicle for diverse forms of late capitalist accumulation, she frames the forms of systemic racism emblematized and anchored by carceral institutions as "the state-sanctioned or extralegal production and exploitation of group-differentiated vulnerability to premature death."[223] In addition to illustrating the way prisons—not only privatized prisons but also public ones—represent multiple opportunities for the production of profit. Gilmore argues that prisons represent a "fix" for the inherent crises of capitalist accumulation. Here, Wilson borrows from and expands a Marxist theorization of the way capitalism necessarily strives to create "fixes" for the inexorable and insoluble contradictions and crises it awakens. As the story so far has perhaps revealed, capitalism is no conspiracy of rich elites, though it can include collusion and cooperation among them. It is, rather, a system driven forward by competition and rivalry between profit-seeking capitalists and corporations. Because of this inherent, relentless competition, and because of the "friction" created everywhere by the resistance of people, capitalism is constantly erupting in major episodic and systemic crises. These crises cannot be permanently resolved without undermining the fundamental nature of the system, and so they must be provisionally "fixed" in various ways. But each fix ultimately displaces the contradiction, which manifests in a new and often seemingly unrelated form elsewhere.[224]

For Wilson, the "prison fix" of mass incarceration not only allows for many lucrative opportunities in staffing, managing, maintaining, and provisioning these institutions, and not only justifies increased spending on policing. It also provides more systemic fixes as well. Key among these is a "fix" for the growing problem of so-called "surplus populations," people who are, on the one hand, dependent on capitalist wages for their survival but, on the other hand, for whose labor power capitalism does not have a sufficiently profitable use.[225] Capitalism, driven by competition between capitalists, creates many circumstances in which workers find themselves unemployed, for instance when competition drives down prices to such an extent that some firms are unable to compete and close shop, or when new technologies arise that mean whole areas of production are retrofitted so as to not require workers in the traditional fashion, or which make previous products redundant. These often lead to periodic episodes of unemployment for some workers and, occasionally, prolonged and generalized periods like recessions or depressions.[226]

Capitalism has always been marked by the problem of surplus populations. It has a structural tendency to disrupt or displace people's non-capitalist ways of life.[227] Colonialism, the European enclosures of common lands, and, increasingly, intensive extractive projects contribute to the trend that sees people often violently removed from the land that sustained them, typically forcing them to migrate for work, often to cities or plantations.[228] As we have seen in the case of South East Asia's palm oil plantations, these displaced workforces can be extremely profitable for those who need a cheapened and exploitable source of labor.[229]

In a previous era, palm oil greased the gears of capitalism's machines of industry and of war. We have seen how it has also provided the kind of commodities that helped entrench racist ideologies of colonialism and bourgeois patriarchy to divide the oppressed of the world against one another. And we have seen how it was literally weaponized to help "contain" anti-capitalist counterinsurgency around the world. Now, in the dormitories of China's booming cities or in the US prison system, palm oil once again facilitates capital's reproduction, but in a new and different way: Rather than reproduce the conditions for the accumulation of capital through industrial manufacturing and state violence, palm oil in these cases, quite literally facilitates the reproduction of the proletarian or sub-proletarian body as surplus flesh.

This facilitates the reproduction of capitalism in a moment when it has so thoroughly stitched itself into the fabric of life and human existence that today, perhaps the majority of human beings on the planet are stripped of their means of subsistence and made dependent on capitalism for their survival.[230] But it is a form of capitalism that increasingly has no economic purpose for them as workers, only as consumers and then only to the extent that they have money or access to debt. And yet some residual moral scruples of the powerful or perhaps the need to maintain a vast "reserve army" of labor means that the surplused are not all simply allowed to starve. They are, for structural reasons, instead sacrificed: incarcerated in death-dealing prisons, consigned to "sacrifice zones" where slow death awaits, allowed to die or actively murdered as they seek to move across borders to escape their fate.

Let us be clear that the only thing "surplus" about surplus populations is that they are made so by the system that cheapens their lives. They are far from "surplus" to their loved ones. Many are the inheritors of invaluable cultural traditions, other cosmologies, and ways of life that offer powerful alternatives to the capitalist order of value that renders them sacrificial. Liberal rhetoric of individualist, market oriented human rights has little to say in the face of such surplussing, which is endemic to the capitalist system liberalism fetishizes and aims to preserve.

A world of surplus people defines some of the key parameters of politics in this "age of human sacrifice." Increasingly xenophobic mainstream political and economic debates circulate around why certain people actually deserve to be surplus and should be allowed to suffer or die. These debates that draw on and renovate the heinous archive of racist tropes that presume some people are fundamentally inferior.[231] As Robbie Shilliam makes clear, the discourse of the "undeserving poor," those who are held to be responsible for their fate has long been weaponized against non-white people in Europe and beyond.[232] Today, fear of the surplus, or fear of becoming surplus is profoundly consequential and frequently animates reactionary fantasies, including the conspiracy theory of the "Great Replacement," which centers on fears of a "white genocide" where "floods" of immigrants to "Western" nations will lead to the demographic evaporation of whiteness and what the proponents characterize as "white culture."[233] But even in non-Western nations far-right and fundamentalist political actors have successfully mobilized fear of surplus-ing to catalyze electoral victory.[234] Here, the image of indifferent hordes of needy others, either attacking head-on or sneaking behind the fortress walls, becomes a common trope, and one that justifies almost any defensive measures. In the name of controlling or containing this surplus flesh, any sacrifice becomes permissible: would-be migrants are drowned in the Mediterranean, made to expire from thirst or exposure on the US-Mexico border, killed by preventable illness in Gaza, starved or made vulnerable to civil war in so-called "failed states," or murdered by drone "for their own good" in neo-imperialist interventions.

# Whose sacrifice?

We began this story with the invasion of the Edo Kingdom in part because of the persistent irony that the British empire essentially justified the mass sacrifice of an entire civilization in the name of intervening to stop human sacrifice. This justification was almost entirely cynical except to the extent it could be used to facilitate the support of the British public for the invasion by trading in racist tropes.

I have sought throughout this text to draw a line from that event through a history of palm oil to the present "age of human sacrifice," a moment when capitalism appears globally triumphant. The notion of sacrifice certainly resonates, as Wendy Brown suggests, in an age of endless austerity when, in contrast to a rhetoric of temporary belt-tightening in the name of future economic growth and abundance, today politicians and business leaders have given up any optimism that sacrifice today will mean anything more than increasing competition, precariousness, and inequality; in other words, more sacrifice.[235]

But is it sacrifice if it is the unintended by-product of the unregulated market? Doesn't sacrifice require it to occur within a context where the death is rendered sacred? Shouldn't the thing or tion practiced these dark rites. What is of greater interest is how stories about the barbaric sacrificial customs of the other are used to hide, mystify or normalize one's own society's forms of human sacrifice, as we saw in the case of the British narratives of the Edo Kingdom's practices.[239]

Perhaps for all societies that practice this horrific ritual, the terrible truth of it is too much to bear, too stark and intimate to face directly. Indeed, often the individuals so sacrificed were cosmologically demoted from even belonging to the same humanity, attributed a status closer to animals and objects.[240] In many cases, human sacrifice was largely the mystified expression of power.[241] Civilizations of all kinds practice episodic sacrifice, especially at moments of crisis. But sacrifice tends to become systemic and regular in societies with pronounced divisions of labor and rigid hierarchies, and often in the name of preserving these power imbalances.[242] Human sacrifice allows elites to murder their rivals, intimidate their vassals and create a spectacle that links them to the power of the gods. They might entrench and preserve their authority by binding it up in a common cosmology which entitles them exclusively—or even requires them—to ritualistically take human life in the name of the common good. Were the blood not spilled, the Gods would be displeased, or would starve, portending apocalyptic calamity for everyone. Framed this way, a life, especially the devalued life of a slave, a vassal, a prisoner of war, or a low-caste non-person, is a small price to pay for providence and prosperity for a whole society.[243] In this sense, human sacrifice is often presented by the elites who practice it as a kind of cosmic insurance: a relatively small, judicious investment made now in the name of averting the potential of future catastrophe.[244]

How different is this rhetoric from that espoused by the defenders of the Benin Punitive Expedition, who argued that the civilizing influence of Christian doctrine and "free trade" justified ethnocidal mass murder? For them, as regrettable as the murder was, it was in the overarching greater interests of civilization as a whole and even of Africans themselves, who, it was said, would ultimately benefit. How different is this from the sermons of the technocratic high priests of racial capitalism today, who do not fail to provide elaborate and mathematically ingenious justifications for why so many millions of people must be sacrificed and rendered "surplused" in the name of the allegedly free market, which, if so appeased, will deliver progress, prosperity, and peace?

The prophets and philosophers of the neoliberal free market ironically frame it as the ultimate transcendence of humanity's barbaric heritage. For Hayek, perhaps the most brilliant exponent of free markets and the intellectual force that is often credited with kickstarting the neoliberal revolution, the governance of society by markets represented the triumph and end point of the Enlightenment, the End of History as Francis Fukayama would later frame it.[245] In this view, because they are composed of the competing intentions of millions of individuals and corporations, markets, when left to their own devices, are essentially a kind of perfectly rational meta-intelligence. Participation in markets, in turn, encourages actors to behave in rational ways, to maximize gain and minimize loss.[246] Superstition, fetishism, and tradition become liabilities that cloud one's ability to study, intuit, and respond to market signals and so compete. Within such a rational system there is no room for the kind of metaphysical "noise" that might be used to legitimate human sacrifice. In the more popular framing of Bill Gates, the friction-free capitalism of free markets would liberate the world from outdated customs and prejudices, much in the same way that nineteenth century British candle and soap advertisers or proponents of the Benin Punitive Expedition insisted that "free trade" (at the end of a gun) would liberate Africans from their barbaric fetishes.[247]

And yet then as now, the belief in the infallible rationality of the market and its idealized rational actor, *homo economicus*, economic man, is itself an example of that thing Europeans came to call the fetish: a thing of this world given magical powers and transcendental status within a particular cosmology. And it is within this cosmology of free markets and *homo economicus* that we need to locate the kinds of human sacrifice we see throughout the history and contemporary material relations of palm oil.

Within this sacrificial cosmology, no single hierophant or high priest needs to wield the fatal blade, and no one need bear witness to the horror. The sacrifice transpires in the clinical anonymity of market relations. An increase in demand for snack food in India triggers a chain of market decisions that see the forced displacement of an indigenous community in West Puapa and, with it, the liquidation of their entire lifeway and cosmology. The anonymous demands of shareholders in a cosmetics firm for greater returns leads to land grabbing by entrepreneurial smallholders or their hyper-exploitation of migrant workers. A subtle shift in policy to encourage markets to turn to biofuels triggers a wave of peatland burning that releases massive amounts of carbon into the atmosphere, contributing to murderous impacts on global human and non-human populations via the vicissitudes of climate change.

The capitalist system has always been one whose broadest patterns are defined not by the collusion of a sacrificial elite, but by competition between capitalist actors at many levels. The invasion of the Edo Kingdom, for instance, was in some senses a conspiracy of like-minded and self-interested traders and colonial officials. But it was pursued by the British not primarily out of wanton cruelty but to satisfy the needs of industry and empire in a moment of heated inter-imperialist rivalry over cheapened labor and resources.[248] The sacrifice of millions of lives and many African kingdoms in this period was incidental, though certainly the devaluation that permitted it was the result of centuries of murderous anti-African and anti-Black ideology.

Today, the millions of people placed on the metaphoric altar of the palm oil industry are not the victims of some sadistic cult but what Randy Martin calls an "empire of indifference," one where everyone and everything is rendered in financialized terms and we are each encouraged to define ourselves and our relation to the world in market terms, adopting the persona of *homo economicus* to manage a lonely world of risks and opportunities.[249]

But the result of all these millions of isolated acts of risk management is a world of chaos. As we saw in the 2008 financial crisis, innumerable acts of highly rational financial activity can and frequently do murmurate into an irrational swarm that ultimately spirals out of control.[250] What emerges is a world of silent bureaucratic financialized violence.[251] The rapid and catastrophic destruction of peatlands by the palm oil industry continues apace not because it is orchestrated by some centralized policy (in fact, most centralized policies claim to work against it). It is driven forward by tens of thousands of entrepreneurial acts, forms of isolated and solipsistic risk management, encouraged by a system in which nearly all are made to participate but beyond anyone's or any government's control. Yet we do not all participate equally or with the same consequences.[252]

This is the state of capitalism today, and why its particular forms of human sacrifice appear to come from everywhere and nowhere at once. It occurs within the context of a broader market cosmology, wherein individual rational acts produce irrational results. The victims are recast as, at best, hapless and unfortunate casualties on the march of economic progress, the benevolent unfolding of the providential market toward the end of history or friction-free capitalism. At worst, the victims are blamed for their failure to embrace the market cosmology and recast themselves as *homo economicus* and compete for their own benefit. In this order of human sacrifice, there need be no physical altar, and no single villainous tyrant, soaked in blood; the altar is the silent ledger, the unaccountable accounting of a financialized, decentralized empire.

In this moment, it should perhaps come as no surprise that we see the resurfacing of demonic fantasies of conspiracy.[253] At the time of writing, the basic tenets of the QAnon conspiracy, which holds that a secret society of economic and political elites and A-list celebrities are running a network of pedophilia and child sacrifice, is said to have so many adherents in the United States alone that it would rank among the nation's top five religions if so classified.[254] Around the world, witch-hunting and witch burnings are on the rise.[255] In both cases, in the face of a generalized, ambient economy of human sacrifice in which we are also deprived of the resources to understand what is happening, we indirectly but persistently, project this heinous cruelty onto individuals, searching for a human source for a metahuman cruelty.

# Whose story?

How does one write a story of palm oil from within that very story? This has been a challenge throughout this book. Both the author and the reader are body-minds that built themselves through the metabolism of palm oil. This book written on a machine whose construction was facilitated by palm oil and is perhaps being read on pages or with ink that include palm oil derivatives. It was published in a world made of palm oil, and within a system that created palm oil as we know it: that alchemical wonder, so foreign to its ancestral virgin form cultivated for millennia in West Africa. It has proven impossible to tell such a story, our story, a story of friction and connection, of profit and sacrifice. The narrative is both too big and too close. In a sense, any story about palm oil is necessarily an autobiography, and, as such, is always generatively flawed, the confession of a subject that can never fully know itself.

At stake for me is the hope that, by striving for a better understanding of palm oil, we can come to recognize our potential as a cooperative species. Today, hundreds of millions of people "cooperate" in the palm oil economy: as growers or consumers, as migrant workers or corporate executives, as any one of a cast of different characters whose proverbial hands have touched the product on its path from seedling to finished product. By cooperation here I do not mean to diminish the vast inequalities and forms of exploitation and unfreedom that define its production and distribution. Rather, I want to reorient our attention towards our entanglements as a cooperative species. Under the forms of racial capitalism that has become triumphant, whose ascendency the story of palm oil helps us map, our cooperation is crystallized in commodities like those thousands of products that are made of or contain or somehow depend on palm oil. It is a sublime web, beyond the scope of our individual imaginations, but we have to try because, as the story of palm oil reveals, if we continue to act as if we are not an interconnected, interdependent cooperative species, we will continue to do catastrophic harm to ourselves and to the planet on which we depend. If we created palm oil and, through it, created our world, what else might we have created? What else might we yet create?

We are, after all, a storytelling species, *homo narrans* as Sylvia Wynter dubs us.[256] In her work, the human species appears as uniquely capable of transforming itself *alchemically* through stories: the narratives we spin, collectively, shape how we reproduce ourselves individually and collectively. Unlike seemingly all other animals, we can radically adjust the values that govern kin selection, the transformation of the material world, and social relationships based on how we narrate our existence to ourselves.

Wynter's core concern is how the global racial capitalist system that emerged from the world-shattering event of the transatlantic slave trade persists in reproducing a global anti-Black cosmology, and how it makes imaginable the superhuman figure of *homo economicus*, forged in the crucible of white supremacy. Today, in what she characterizes as a unique, world-historical catastrophe, the white supremacist idealized story of "economic man" has become paradigmatic and reproduces the world in its image. Wynter's approach helps us explain why anti-Blackness recurs again and again in this story, from the justification for the sacking of the Edo Kingdom to the scene of racialized class war in the American prison gulag.

But I am also drawn to the tantalizing possibilities of Wynter's broader characterization of humanity as an *alchemical*, storytelling species. I have argued throughout this book, in keeping with Gill and Taussig's invitation, that we are all alchemically palm oil now: it is part of all of us, metabolized into our bodies, saturating our skin, instilled in the tools and commodities we use, haunting the economy of which we are a part. What stories can and should our hybrid, alchemical species tell about itself to transform itself away from the cosmology of the market and its order of sacrifice?[257]

The stories we have been encouraged to tell about palm oil are not sufficient to this task. We cannot allow our imaginations to be confined to the orbit of the figure of the enlightened consumer, who is simply one facet of *homo economicus*. Even if massive consumer and political pressure were to be brought to bear, I am deeply skeptical of claims that the palm industry can be reformed in a substantial way. We have already seen how adept the industry is at evading regulation and capturing government interests. Voluntary schemes to curtail the worst abuses have been largely unsuccessful within the context of a constantly expanding industry keen to meet international demand.[258] And even if they could leverage pressure in one nation or jurisdiction, there are many others vying for a share of the market: if regulation were enforced in Malaysia and Indonesia, which today are responsible for 80% of exports, there are significant operations in South East Asia, Latin America, West Africa, and elsewhere in the tropical regions of Asia.

Meanwhile, a number of the palm oil industry's self-serving claims have an element of truth to them. *E. guineensis* is, by most standards, the most efficient oil-producing plant in terms of gross acreage, and it also thrives in conditions unsuitable to many other crops.[259] The global population is growing, and with it the demand for cheap food and fat, especially as this population growth is being accompanied by trends toward urbanization and displacement that make people dependent on imported foodstuffs. If, magically, palm oil were to vanish tomorrow and other fats were to take its place, things wouldn't necessarily be better: the environmental, labor, and humanitarian track records of other intensive plant oils, like soy, are not without their horrors.[260] As more and more chemical techniques are developed, palm oil, which is at least nominally derived from a renewable source, could replace petroleum in some products. Further, for all the harms it has caused, palm oil has undeniably buoyed (highly unequal) economic "development" in South East Asia in ways other cash crops have failed to do elsewhere.[261]

Ultimately, the problem is not palm oil per se, but the context and methods in which it is manufactured and the functions it serves within a broader capitalist world system. Reports indicate that smallholders who cultivate oil palm as part of a diversified array of crops can rival the efficiency per acre of huge plantations, but rarely match the price, which depends on the artificial cheapening of plantation labor and exploitative economies of scale.[262] It is not inconceivable that, in a better, fairer arrangement of global affairs, palm oil might be one particular ingredient in some globally traded commodities, a supplement to diets rich in more locally sourced and wholesome oils. But it cannot be sustained, as it is today, as the artificially cheapened fat of the world's poor, whose function is increasingly to feed the poorest people on the planet otherwise left to die by the vicissitudes of global capitalism.

In the end, the solutions to the violence and exploitation of the palm oil industry will be determined by its workers and the communities it affects. A consortium of researchers and campaigners for grassroots organizations have recently developed a sketch of a roadmap for a "just transition" in which workers neither simply defend themselves from the worst excesses of the industry, nor ask governments to fix the problem. Rather,

porting food security, health and the well-being of the local population." Importantly, such "social and ecological functions cannot all be measured in monetary terms," defying the relentless logic of the market and grounding such a transition in different cosmologies. Such a plan would necessarily drive up the cost of palm oil on international markets, but that is not necessarily bad for workers or consumers: as we have already seen, cheap palm oil has in many places been a curse.

Because it focuses on the insights and struggles of affected communities, Indigenous people, and workers, this scheme is to my mind highly promising, especially in a world where discussion of "solutions" tends to be dominated by negotiations by corporations, governments, and occasionally large NGOs, who are typically considered the only legitimate or reliable stakeholders. In official negotiations, the sacrificial cosmology of the market is taken as a given, something to be worked around or accommodated rather than fundamentally rejected and abolished. At the root of the just transition proposal, is an important framework that might inspire consideration from those of us not directly involved in those regions.

In the first place, it would call upon those of us who depend on palm oil and similar cheapened fruits of exploitation and extraction to radically transform our ways of life and material culture. We would also need to consider how we would act in solidarity with affected communities as they came together to demand a just transition in the face of an extremely powerful consortium of landholders, corporations, and governments who have every reason to seek to retain and expand the current system.

But more is at stake in the Just Transition proposal. As the authors note, their proposal for a just transition is rooted in the common experience of alienation from land. This begs the broader question of a global solidarity that is also rooted in practices of dis-alienation and reconnection, to land, to one another, and to the cooperative enterprise of the species of which we are a part.

The question that arises is how this same sensibility might animate a global revolution? Because the problem of palm oil is a global one and has encrypted within it the question of what humanity will become in a moment when we can no longer afford to act individually or even regionally. To borrow a question from prison abolitionists: what would it take to create a world which no longer perceived itself to need industrially produced palm oil? What steps would need to be taken now and in each locale to begin a transition that would reduce and eliminate our dependence on this cheapened substance? What would it take to create not only the policies and governance structures but the ways of life that would eliminate the violent capitalist category of "cheapness" altogether? What would it mean to enhance the scope and sensitivity of our collective imagination to envision a different habit of relation, or many habits of relation, working in solidarity? What would it mean to consider the kinds of alliances it would take to rebel against those forces of global capitalism that have everything to gain from the perpetuation and expansion of the current order, not only the corporations pushing palm oil and consumer products, and the governments that support them, but their brethren in other sectors as well?

In an age of crisis built on centuries of division and inequality, it has become unfashionable to speak of the need for a global revolution. And yet it seems there is no other option. But for that revolution to truly replace the old order of neocolonial racial capitalism, it will need to be grounded in the kind of thinking and infrastructure of solidarity for which the Just Transition plan calls. We are a species that has created a world of sacrifice and that has been created by a world of sacrifice, a world of separation, instrumentalization, and disposability. A global revolution will require becoming something alchemically different, transforming ourselves not only in terms of how we imagine and cooperate but in the very way we, collectively, cooperatively compose and recompose our bodies, minds, and societies.

VĀG
ABO
NDS

# Acknowledgments

A great many people, directly or indirectly, offered their insights, advice, editorial feedback, and reflections in ways that made this book better. They include: Phanuel Antwi, Chris Arsenault, Dan Hicks, Nehikhare Patrick Igbinijesu, Ben Evans James, Megan Kinch, Aris Komporozos-Athanasiou, Leigh Claire La Berge, Mao Mollona, Tina Munroe, Christian Nagler, Simon Orpana, Sina Ribak, David Shulman, Cassie Thornton, Ezra Winton, and Anna-Esther Younes. I am grateful for the opportunity to share this work at The ReImagining Value Action Lab's summer institute, The Moos. Garden Residency, The Sandberg Institute, and via a public lecture series at the Halifax Public Library.

VĀG
ABO
NDS

# Notes

1.  Jocelyn C. Zuckerman, *Planet Palm: How Palm Oil Ended Up in Everything—and Endangered the World* (London: Hurst, 2021).

2.  Haiven, Max, *Cultures of Financialization: Fictitious Capital in Popular Culture and Everyday Life* (London and New York: Palgrave Macmillan, 2014).

3.  Haiven, Max, *Revenge Capitalism: The Ghosts of Empire, the Demons of Capital, and the Settling of Unpayable Debts* (London and New York: Pluto, 2020).

4.  Ruth Wilson Gilmore, *Golden Gulag: Prisons, Surplus, Crisis, and Opposition in Globalizing California* (Berkeley, CA: University of California Press, 2007), p. 244.

5.  Jonathan E. Robins, *Oil Palm: A Global History* (The University of North Carolina Press: Chapel Hill, NC, 2021), Chapter 2.

6.  See Robins, *Oil Palm*, Chapter 2; Lynn, Martin. *Commerce and economic change in West Africa: The palm oil trade in the nineteenth century* (Cambridge and New York, 1997), p. 2; On the contemporary importance of dendê palm oil in Afro-Brazilian spirituality and culture, see Zuckerman, *Planet Palm*, pp. 56–9.

7.  Zuckerman, *Planet Palm*, Chapter 2.

8.  "Alternative Names to Palm Oil." Orangutang Alliance, October 2021, https://

orangutanalliance.org/whats-the-issue/
alternative-names-for-palm-oil/.

9.     Renato J. Orsato, Stewart R. Clegg, and
       Horacio Falcão, "The Political Ecology of
       Palm Oil Production," *Journal of Change Man-
       agement* 13, no. 4 (2013): 444–59.

10.    Oliver Pye, "A Plantation Precariat: Fragmen-
       tation and Organizing Potential in the Palm
       Oil Global Production Network," *Development
       and Change* 48, no. 5 (2017): 942–64.

11.    See Zuckerman, *Planet Palm*.

12.    Robins, *Oil Palm*, pp. 268–9.

13.    Ibid., p. 283.

14.    Zuckerman, *Planet Palm*, pp. 6–7.

15.    See https://ourworldindata.org/land-use.

16.    Nnoko-Mewanu,  Juliana,  "'Why  Our
       Land?' Oil Palm Expansion in Indonesia Risks
       Peatlands and Livelihoods" (London, UK:
       Human Rights Watch, June 2021). https://
       hrw.org/report/2021/06/03/why-our-land/
       oil-palm-expansion-indonesia-risks-peatlands-
       and-livelihoods.

17.    Robins, *Oil Palm*, Chapter 7.

18.    Dauvergne, Peter, "The Global Politics of the
       Business of 'Sustainable' Palm Oil," *Global
       Environmental Politics* 18, no. 2 (2018): 37.

19.    See, for example, Chao, Sophie, "Can There
       Be Justice Here? Indigenous Perspectives from
       the West Papuan Oil Palm Frontier," *borderlands*
       20, no. 1 (2021).

20.    Simryn Gill and Michael Taussig, *Becoming
       Palm* (Berlin: Sternberg, 2017).

21.    See Robins, *Oil Palm*, pp. 288–91.

22.    Pye, Oliver, "Commodifying Sustainability:
       Development, Nature and Politics in the Palm

Oil Industry," *World Development* 121 (2019): 218–28.

23. See Watts, Joseph, Oliver Sheehan, Quentin D. Atkinson, Joseph Bulbulia, and Russell D. Gray, "Ritual Human Sacrifice Promoted and Sustained the Evolution of Stratified Societies," *Nature* 532, no. 7598 (2016): 228–31.

24. Bichler, Shimshon, and Jonathan Nitzan, "Capital as Power: Toward a New Cosmology of Capitalism," *The Bichler and Nitzan Archives*, 2010. http://bnarchives.yorku.ca/285/.

25. On the particular (violent) arrogance and solipsism of the modern capitalist cosmology relative to others, see Graeber, David and David Wengrow, *The Dawn of Everything: A New History of Humanity* (New York: Penguin, 2021).

26. Wynter, Sylvia, and Katherine McKittrick, "Unparalleled Catastrophe for Our Species?: Or, to Give Humanness a Different Future: Conversations," In *Sylvia Wynter: On Being Human as Praxis*, edited by Katherine McKittrick (Durham, NC and London: Duke University Press, 2015).

27. Paddy Docherty, *Blood and Bronze: The British Empire and the Sack of Benin* (Hurst: London and New York, 2021).

28. Michael Gibson-Light, "Ramen Politics: Informal Money and Logics of Resistance in the Contemporary American Prison," *Qualitative Sociology* 41, no. 2 (2018): 199–220.

29. Data retrieved January 6, 2022, from https://boxofficemojo.com/chart/top_lifetime_gross/?area=XWW.

30. See Docherty, *Blood and Bronze*.

31. Hicks, Dan, *The Brutish Museums: The Benin Bronzes, Colonial Violence and Cultural Restitution* (London and New York: Pluto, 2020).

32. Aremu, Johnson Olaosebikan and Michael Ediagbonya, "Trade and Religion in British-Benin Relations, 1553–1897," *Global Journal of Social Sciences Studies* 4, no. 2 (2018): 78–90; Docherty, *Blood and Bronze.*

33. See Hicks, *The Brutish Museum*; Savoy, Bénédicte. *Afrikas Kampf Um Seine Kunst Geschichte Einer Postkolonialen Niederlage* (München: C. H. Beck, 2021).

34. Ndikung, Bonaventure Soh Bejeng, "South Remembers: Those Who Are Dead Are Not Ever Gone," *South as a State of Mind* 10 (2018). https://southasastateofmind.com/south-remembers-those-who-are-dead-are-not-ever-gone/.

35. Matory, J. Lorand, *The Fetish Revisited: Marx, Freud and the Gods Black People Make* (Durham and London: Duke University Press, 2018).

36. On the imperial and capitalist uses of vengeance, see Haiven, *Revenge Capitalism*, Chapter 1.

37. Robins, *Oil Palm*, pp. 45–8. Martin Lynn, *Commerce and Economic Change in West Africa*, p. 2; Zuckerman, *Planet Palm*, p. 47.

38. Lynn, *Commerce and Economic Change in West Africa*, p. 3.

39. Statistics are drawn from Lynn, *Commerce and Economic Change in West Africa* p. 3 and p. 49. The UK palm oil imports from Africa in 1897 were 1,262,933 hundredweight (CWT – i.e. 1/20 of an imperial ton). At roughly the same period, the workdays required to produce a ton for export-ready palm oil were: 420 for "soft" (more refined) palm oil and 132 for "hard,"

though Robins (*Oil Palm*, pp. 26–7) has reason to doubt these numbers. Soft oil was significantly more in demand for use as machine oil, some kinds of soap, margarine, and flux for the manufacture of tin. These figures do not include the trade in raw palm kernels. The measurement of a "workday" at the time in Britain presumed a ten-hour day. However, it is highly likely that African palm oil workers did not enjoy this luxury. Robins provides a variety of other estimates in *Oil Palm*, Chapter 3.

40. Robins, *Oil Palm*, pp. 68–9.

41. Robins, *Oil Palm*, pp. 117–8.

42. Teltscher, Kate, *Palace of Palms: Tropic Dreams and the Making of Kew* (London: Picador, 2020).

43. Rodney, Walter, *How Europe Underdeveloped Africa*, new edition (London and New York: Verso, 2018), Chapter 5.

44. Robins, *Oil Palm*, Chapter 3.

45. Robins, *Oil Palm*, pp. 92–103.

46. Docherty, *Blood and Bronze*, pp. 12, 30.

47. Pietz, William, "The Problem of the Fetish," *RES: Anthropology and Aesthetics* 9 (1985): 5–17.

48. Nassau, Robert Hamill, "Fetishism, a Government," *Bulletin of the American Geographical Society* 33, no. 4 (1901): 305.

49. On this imperial weaponization of liberalism and the rhetoric of "free trade," see Lowe, Lisa, *The Intimacies of Four Continents* (Durham NC and London: Duke University Press, 2015).

50. Docherty, *Blood and Bronze*.

51. For a gloss of coverage of sacrifices witnessed by British officials and the British press, see Docherty, *Blood and Bronze*, pp. 183–9.

52. See, for instance, Kaplan, Flora Edouwaye S., "Understanding Sacrifice and Sanctity in Benin Indigenous Religion, Nigeria," in *Beyond Primitivism: Indigenous Religious Traditions and Modernity*, edited by Jacob K. Olupona (London and New York: Routledge, 2003), pp. 181–99.

53. See Watts et al., "Ritual Human Sacrifice Promoted and Sustained the Evolution of Stratified Societies."

54. Pietz, William, "The Spirit of Civilization: Blood Sacrifice and Monetary Debt," *Res: Anthropology and Aesthetics* 28 (1995): 23–38.

55. Todorov, Tvetan, *The Conquest of America: The Question of the Other*, trans. Richard Howard (New York: Harper Torchbooks, 1984).

56. See Docherty, *Blood and Bronze*, pp. 106–8.

57. Docherty, *Blood and Bronze*, p. 197.

58. See Ukaogo, Victor, "The Emergence of Social Movements in Southwestern Nigeria, with Specific Reference to Esan and Their Neighbours (Benin Province): C 1900–1960 A.D.," 2013. https://researchgate.net/publication/338686480_A_Diminishing_Past_A_Rescued_Future_Essays_on_the_Peoples_Traditions_and_Culture_of_the_Esan_of_Southern_Nigeria.

59. Robins, *Oil Palm*, Chapters 6–7.

60. Robins, *Oil Palm*, pp. 186–7.

61. Robbins, *Oil Palm*, Chapter 3.

62. Tsing, Anna Lowenhaupt, *Friction: An Ethnography of Global Connection* (Princeton and London: Princeton University Press, 2004).

63. Gates, Bill, *The Road Ahead* (New York: Viking, 1995); See also McGoey, Linsey, *No Such Thing as a Free Gift: The Gates Foundation and the Price*

*of Philanthropy* (London and New York: Verso, 2016).

64. Lynn, *Commerce and Economic Change in West Africa.*

65. For a survey of the chemistry sector in Liverpool in the nineteenth century, see Reed, Peter, *Entrepreneurial Ventures in Chemistry: The Muspratts of Liverpool, 1793–1934* (London and New York: Routledge, 2015).

66. Lynn, *Commerce and Economic Change in West Africa.*

67. There remains a scholarly debate on why the price of palm oil fell in this period. For a summary of positions, see Robins, *Oil Palm,* Chapter 4.

68. On these debates, see Hicks, *The Brutish Museums.*

69. Luxemburg, Rosa, *The Accumulation of Capital,* trans. Agnes Schwarzschild (London and New York: Routledge, 2003).

70. This century-long process is detailed by Robins, *Oil Palm,* Chapters 3, 4, and 6.

71. Robins, *Oil Palm,* pp. 112–5.

72. Robins, *Oil Palm,* p. 124.

73. Ranciere, Jacques, *Proletarian Nights: The Workers' Dream in Nineteenth-Century France,* trans. John Drury (London and New York: Verso, 2012).

74. See Crary, Jonathan, *24/7: Late Capitalism and the Ends of Sleep* (London and New York: Verso, 2013); Melbin, Murray, *Night as Frontier: Colonizing the World After Dark* (New York: The Free Press, 1987).

75. See Teltscher, *Palace of Palms.*

76. Robins, *Oil Palm,* pp. 115–16.

77. McClintock, Anne, *Imperial Leather Race, Gender, and Sexuality in the Colonial Contest* (London and New York: Routledge, 1995), Chapter 2.

78. Lynn, *Commerce and Economic Change in West Africa*, pp. 29–30.

79. Robins, *Oil Palm*, pp.105–16.

80. Lynn, *Commerce and Economic Change in West Africa*, p. 85.

81. Zuckerman, *Planet Palm*, Chapter 3.

82. Robins, *Oil Palm*, p. 196.

83. Patel, Raj and Jason W. Moore, *The History of the World in Seven Cheap Things* (Berkely, CA: University of California Press, 2017).

84. Wolford, Wendy, "The Plantationocene: A Lusotropical Contribution to the Theory," *Annals of the American Association of Geographers*, 2021, 1–18; See also Li, Tania Murray and Pujo Semedi, *Plantation Life: Corporate Occupation in Indonesia's Oil Palm Zone* (Durham NC and London: Duke University Press, 2021).

85. See Robins, *Oil Palm*, Chapter 5.

86. McClintock, *Imperial Leather*, pp. 207–13.

87. Zuckerman, *Planet Palm*, pp. 61–2.

88. Rodney, *How Europe Underdeveloped Africa*, Chapter 5.

89. Zuckerman, *Planet Palm*, pp. 61–4.

90. Federici, Silvia, *Caliban and the Witch: Women, The Body and Primitive Accumulation* (Brooklyn NY: Autonomedia, 2004).

91. McClintock, *Imperial Leather*.

92. Hobsbawm, Eric, *Industry and Empire: The Birth of the Industrial Revolution*, ed., Chris Wrigley (New York: The New Press, 1999), Chapters 4 and 8.

93. Robins, *Oil Palm*, pp. 109–10.

94. Bremmer, Jan N., ed., *The Strange World of Human Sacrifice* (Leuven: Peeters, 2007).

95. McClintock, *Imperial Leather*, 207–31.

96. Nelson, Anitra, *Marx's Concept of Money* (London and New York: Routledge, 1999), Chapter 7.

97. Freud, Sigmund, "Fetishism," *International Journal of Psychoanalysis* 9, no. 2 (1928): 161–6.

98. See Žižek, Slavoj, *The Sublime Object of Ideology* (London and New York: Verso, 1989).

99. See da Silva, Denise Ferreira, *Towards a Global Idea of Race* (Minneapolis: University of Minnesota Press, 2007).

100. Matory, *The Fetish Revisited*.

101. Pietz, "The Problem of the Fetish."

102. Robins, *Oil Palm*, p. 306.

103. On how such priorities get set, and their consequences, see Krause, Monika and Katherine Robinson, "Charismatic Species and beyond: How Cultural Schemas and Organisational Routines Shape Conservation," *Conservation & Society* 15, no. 3 (2017): 313–21.

104. Pye, "Commodifying Sustainability."

105. Robins, *Oil Palm*, Chapter 4.

106. Zuckerman, *Planet Palm*, pp. 88–9.

107. On the importance of gun oil, see Satia, Priya, *Empire of Guns* (Stanford CA and London: Stanford University Press, 2019).

108. Docherty, *Blood and Bronze*.

109. See Reed, 2015, *Entrepreneurial Ventures in Chemistry*.

110. See Wagner, Kim A., *The Skull of Alum Bheg: The Life and Death of a Rebel of 1857* (Oxford and New York: Oxford University Press, 2017).

111. Fant, Kenne, *Alfred Nobel: A Biography*, trans. Marianne Ruth (New York: Arcade, 1993).

112. Lynn, *Commerce and economic change in West Africa*, p. 118; Zuckerman, *Planet Palm*, p. 75.

113. Jensen, Richard, "Daggers, Rifles and Dynamite: Anarchist Terrorism in Nineteenth Century Europe," *Terrorism and Political Violence* 16, no. 1 (2010): 116–53; Werrett, Simon, "The Science of Destruction: Terrorism and Technology in the Nineteenth Century," in *The Oxford Handbook of the History of Terrorism*, eds., Carola Dietze and Claudia Verhoeven (Oxford and New York: Oxford University Press, 2022).

114. Arboleda, Martín, *Planetary Mine: Territories of Extraction under Late Capitalism* (London and New York: Verso, 2020).

115. Karuka, Manu, *Empire's Tracks: Indigenous Nations, Chinese Workers, and the Transcontinental Railroad* (Berkeley CA and London: University of California Press, 2019).

116. Fromm, Erich, *The Anatomy of Human Destructiveness* (New York: Henry Holt, 1973), p. 207.

117. Richardson, John, "A Different Guernica," *New York Review of Books*, (May 12, 2016), 63, no. 8: 4–6; Koussiaki, Fotini. "The Influence of Non-Traditional Art Materials on the Paintings of Pablo Picasso," in *Papers Presented at the Thirtieth Annual Meeting of The American Institute for Conservation of Historic and Artistic Work*, edited by Helen Mar Parkin, 2002, pp. 37–48.

118. Leighten, Patricia, "The White Peril and L'Art Negre: Picasso, Primitivism, and Anticolonialism," *The Art Bulletin* 72, no. 4 (1990): 609–30.

119. Clifford, James, "On Ethnographic Surrealism," *Comparative Studies in Society and History* 23, no. 4 (1981): 539–64.

120. Blier, Suzanne Preston, *Picasso's Demoiselles: The Untold Origins of a Modern Masterpiece* (Durham NC and London: Duke University Press, 2019).

121. Hicks, *The Brutish Museum*.

122. See Haiven, Max, *Art After Money, Money After Art: Creative Strategies Against Financialization* (London and New York: Pluto, 2018).

123. Neer, Robert M., *Napalm: An American Biography* (Cambridge MA and London: Belknap, 2015).

124. Robins, *Oil Palm*, pp. 125–7.

125. Robins, *Oil Palm*, p. 128.

126. Simon Naylor, "Spacing the Can: Empire, Modernity, and the Globalisation of Food," *Environment and Planning A: Economy and Space* 32, no. 9 (September 2000): 1625–1639.

127. Robins, *Oil Palm*, pp. 120–22.

128. Lowe, Kate and Eugene McLaughlin, "'Caution! The Bread Is Poisoned': The Hong Kong Mass Poisoning of January 1857," *The Journal of Imperial and Commonwealth History* 43, no. 2 (2015): 189–209.

129. Chappelle, Frankie, "The Poison of Empire," *The Royal Society Blog*, December 2020, https://royalsociety.org/blog/2020/12/the-poison-of-empire/.

130. Docherty, *Blood and Bronze*, p. 163.

131. Naylor, "Spacing the Can," 1635–6.

132. Möckel, Benjamin, "The Material Culture of Human Rights. Consumer Products, Boycotts and the Transformation of Human Rights Activism in the 1970s and 1980s," *International Journal for History, Culture and Modernity* 6, no. 1 (2018): 76–104.

133. Zuckerman, *Planet Palm*, p. 67.

134. Lynn, Commerce and Economic Change in West Africa, p. 118; Zuckerman, *Planet Palm*, p. 75.

135. Zuckerman, *Planet Palm*, Chapter 9.

136. Basu, Sanjay, Kim S Babiarz, Shah Ebrahim, Sukumar Vellakkal, David Stuckler, and Jeremy D Goldhaber-Fiebert, "Palm Oil Taxes and Cardiovascular Disease Mortality in India: Economic-Epidemiologic Model," BMJ : *British Medical Journal* 347, no. oct22 3 (2013): f6048.

137. Patel and Moore, *The History of the World in Seven Cheap Things*.

138. Zuckerman, *Planet Palm*, pp. 168–9.

139. Robins, *Oil Palm*, pp. 285–6

140. Pearce, Fred, "UK Animal Feed Helping to Destroy Asian Rainforest, Study Shows," *Guardian*, May 2011, https://theguardian.com/environment/2011/may/09/pet-food-asian-rainforest.

141. Mancini, Annamaria, et al., "Biological and Nutritional Properties of Palm Oil and Palmitic Acid: Effects on Health," *Molecules* 20, no. 9 (2015): 17339–17361.

142. Zuckerman, *Planet Palm*, Chapter 7; Basu et al., "Palm Oil Taxes and Cardiovascular Disease Mortality in India."

143. Chen, Brian K, et al., "Multi-Country Analysis of Palm Oil Consumption and Cardiovascular Disease Mortality for Countries at Different Stages of Economic Development: 1980–1997," *Globalization and Health* 7, no. 1 (2011): 45.

144. Capecchi, Stefania, Mario Amato, Valeria Sodano, and Fabio Verneau, "Understanding Beliefs and Concerns towards Palm Oil:

Empirical Evidence and Policy Implications." *Food Policy* 89 (2019): 101785.

145. Robins, *Oil Palm*, p. 341.

146. Ibid., pp. 339–40.

147. Stoler, Ann Laura, *Capitalism and Confrontation in Sumatra's Plantation Belt, 1870–1979*, Second Edition (Ann Arbor MI: University of Michigan Press, 1985).

148. Robins, *Oil Palm*, Chapter 7.

149. Bakan, Joel, *The Corporation: The Pathological Pursuit of Profit and Power* (New York: Free Press, 2004).

150. Stoler, *Capitalism and Confrontation*.

151. Robins, *Oil Palm*, Chapter 7.

152. Ibid., p. 229 and Chapter 8.

153. Ibid., Chapter 8.

154. Zuckerman, *Planet Palm*, pp. 95–102.

155. Ibid., pp. 113, 186.

156. Dauvergne, "The Global Politics of the Business of 'Sustainable' Palm Oil."; Zuckerman, *Planet Palm*, Chapter 9; Schneider, Victoria, "How the Legacy of Colonialism Built a Palm Oil Empire," *Mongabay*, June 2020. https://news.mongabay.com/2020/06/how-the-legacy-of-colonialism-built-a-palm-oil-empire/.

157. Dauvergne, Peter, "The Global Politics of the Business of 'Sustainable' Palm Oil," *Global Environmental Politics* 18, no. 2 (2018): 34–52.

158. Robins, *Oil Palm*, Chapter 10.

159. Damiani, Sandra, et al., "'All That's Left Is Bare Land and Sky': Palm Oil Culture and Socioenvironmental Impacts on a Tembé Indigenous Territory in the Brazilian Amazon," *Ambiente & Sociedade* 23 (2020): e00492.

160. Potter, Lesley, "Colombia's Oil Palm Development in Times of War and 'Peace': Myths,

Enablers and the Disparate Realities of Land Control," *Journal of Rural Studies* 78 (2020): 491–502.

161. Li and Semedi, *Plantation Life*.

162. Pye, Oliver, "Agrarian Marxism and the Proletariat: A Palm Oil Manifesto." *The Journal of Peasant Studies*, 2019, 1–20.

163. "The Great Palm Oil Scandal: Labor Abuses behind Big Brand Names" (London: Amnesty International, November 2016), pp. 3–4.

164. "Empty Assurances" (Bogor, Indonesia and Washington DC: Sawit Watch and the International Labor Rights Forum, November 2013), https://laborrights.org/sites/default/files/publications-and-resources/Empty%20Assurances.pdf.

165. "U.S. Blocks Palm Oil Imports from Malaysia's Sime Darby over Forced Labor Allegations," *Reuters*, December 2020, https://reuters.com/business/energy/us-blocks-palm-oil-imports-malaysias-sime-darby-over-forced-labor-allegations-2020-12-31/.

166. Pattisson, Pete, "Malaysian Prisoners May Face 'Forced Labor' on Palm Oil Plantations." *Guardian*, September 2020, https://the guardian.com/global-development/2020/sep/16/malaysian-prisoners-may-face-forced-labor-on-palm-oil-plantations.

167. Klawitter, Nils, "The Dirty Business of Palm Oil," *Der Spiegel*, February 2014. https://spiegel.de/international/world/indonesian-villagers-driven-from-villages-in-palm-oil-land-theft-a-967198.html; Lustgarten, Abrahm. "Palm Oil Was Supposed to Help Save the Planet. Instead It Unleashed a Catastrophe." *The New York Times*

*Magazine*, November 2018, https://nytimes. com/2018/11/20/magazine/palm-oil-borneo-climate-catastrophe.html.

168. Dauvergne, "The Global Politics of the Business of 'Sustainable' Palm Oil."

169. Mason, Margie, and Robin McDowell, "Rape, Abuses in Oil Fields Linked to Top Beauty Brands," *Associated Press*, November 2020, https://apnews.com/article/palm-oil-abuse-investigation-cosmetics-2a209d60c42bf0e8fcc 6f8ea6daa11c7.

170. Amnesty International, "The Great Palm Oil Scandal."

171. Pye, Oliver, and Jayati Bhattacharya, eds., *The Palm Oil Controversy in Southeast Asia: A Transnational Perspective* (Singapore: Institute for South East Asian Studies, 2013).

172. Pye, "Commodifying Sustainability, 220.

173. Zuckerman, *Planet Palm*, Chapter 8.

174. Meijaard, Erik, Thomas M. Brooks, Kimberly M. Carlson, Eleanor M. Slade, John Garcia-Ulloa, David L. A. Gaveau, Janice Ser Huay Lee, et al., "The Environmental Impacts of Palm Oil in Context," *Nature Plants* 6, no. 12 (2020): 1418–1426.

175. Meijaard, Erik, et al., "Oil Palm and Biodiversity: A Situation Analysis" (Gland, Switzerland: International Union for Conservation of Nature Oil Palm Task Force, 2018), https://portals.iucn.org/library/node/47753.

176. Meijaard et al., "The Environmental Impacts of Palm Oil in Context."

177. Purwestri, Ratna C., et al., "From Growing Food to Growing Cash: Understanding the Drivers of Food Choice in the Context of Rapid Agrarian Change in Indonesia," *CIFOR*

*Infobrief* (Bogor, Indonesia: Center for International Forestry Research, 2019), https://cifor.org/knowledge/publication/7360/.

178. Pye, Oliver, "A Plantation Precariat."

179. Zuckerman, *Planet Palm*, Chapter 8.

180. Lustgarten, "Palm Oil Was Supposed to Help Save the Planet."

181. McDonald, Sharyn, "Managing Issues through Cross-Sector Collaboration: Unilever and Greenpeace," in *Crisis Communication in a Digital World*, eds., Mark Sheehan and Deirdre Quinn-Allan (Cambridge and New York: Cambridge University Press, 2015), 80–91.

182. Zuckerman, *Planet Palm*, Chapters 9–10.

183. Lustgarten, "Palm Oil Was Supposed to Help Save the Planet."

184. Gottwald, Eric, "Certifying Exploitation: Why 'Sustainable' Palm Oil Production Is Failing Workers," *New Labor Forum* 27, no. 2 (2018): 74–82.

185. Robins, *Oil Palm*, pp. 342–5.

186. Genoud, Christelle, "Access to Land and the Round Table on Sustainable Palm Oil in Colombia," *Globalizations* 18, no. 3 (2020): 1–18.

187. Pye, Oliver, "Commodifying Sustainability: Development, Nature and Politics in the Palm Oil Industry," *World Development* 121 (2019): 218–28.

188. Dauvergne, Peter, "The Global Politics of the Business of 'Sustainable' Palm Oil," *Global Environmental Politics* 18, no. 2 (2018): 34–52.

189. Zuckerman, *Planet Palm*, Chapter 8.

190. Wicke, Janis, "Sustainable Palm Oil or Certified Dispossession? NGOs within Scalar Struggles over the RSPO Private Governance

Standard," *Bioeconomy & Inequalities Working Paper Series* 8 (2019), https://bioinequalities. uni-jena.de/sozbemedia/WorkingPaper8.pdf.

191. Zuckerman, *Planet Palm*, Chapter 10; Morales, Juan David López, "Colombia Has Signed a Peace Agreement, so Why Are Trade Unionists Still Being Threatened and Murdered?" *Equal Times*, October 2021, https://equaltimes. org/loomber-has-signed-a-peace?lang=en; Pearce, Fred, "Murder in Malaysia: How Protecting Native Forests Cost an Activist His Life," *The Guardian*, April 2017, https://the-guardian.com/environment/2017/mar/24/ in-malaysia-how-protecting-native-forests-cost-an-activist-his-life.

192. Gerber, Julien-François, "An Overview of Resistance against Industrial Tree Plantations in the Global South," *Economic and Political Weekly* 45, no. 41 (2010): 30–34, https://jstor. org/stable/25742174.

193. Cuffe, Sandra, "Guatemala's Growing Palm Oil Industry Fuels Indigenous Land Fight," *Al Jazeera*, October 2021, https://aljazeera. com/news/2021/10/15/loomberg-growing-palm-oil-industry-fuels-indigenous-land-fight; Jong, Hans Nicholas, "Papua Tribe Moves to Block Clearing of Its Ancestral Forest for Palm Oil," *Mongabay*, January 2022, https:// news.mongabay.com/2021/01/papua-tribe-moves-to-block-clearing-of-its-ancestral-forest-for-palm-oil/; Ionova, Ana, "New Palm Oil Frontier Sparks Scramble for Land in the Brazilian Amazon," *Mongabay*, April 2021, https://news.mongabay.com/2021/04/ new-palm-oil-frontier-sparks-scramble-for-land-in-the-brazilian-amazon/.

194. Serrano, Ángela, "Oil Palm Workers Confront a Fatal Blow Against Unions in Colombia," Collective of Agrarian Scholar-Activists from the South, April 2021. https://casasouth.org/oil-palm-workers-confront-a-fatal-blow-against-unions-in-colombia/; Pye, Oliver, Ramlah Daud, Kartika Manurung, and Saurlin Siagan, "Workers in the Palm Oil Industry: Exploitation, Resistance and Trans-national Solidarity" (Cologne: Stiftung Asienhaus, 2016), https://asienhaus.de/archiv/user_upload/Palm_Oil_Workers_-_Exploitation__Resistance_and_Transnational_Solidarity.pdf.

195. "Resistance Against Industrial Oil Palm Plantations in West and Central Africa," *World Rainforest Movement Bulletin*, no. 254, (February 2021), https://wrm.org.uy/wp-content/uploads/2021/03/Boletin-254_ENG.pdf.

196. Robins, *Oil Palm*, pp. 352–3.

197. Robles, Pablo, et al., "The World's Addiction to Palm Oil Is Only Getting Worse," *Bloomberg*, November 2021. https://bloomberg.com/graphics/2021-palm-oil-deforestation-climate-change/.

198. Robins, *Oil Palm*, pp. 339–40; Zuckerman, *Planet Palm*, pp. 191–3.

199. See Robins, *Oil Palm*, pp. 333–6.

200. Lynn, *Commerce and Economic Change in West Africa*.

201. Robins, *Oil Palm*, pp. 70–73.

202. See, for example, "Palm Oil's Role in Feeding the World," Golden Agri-Resources, September 2018. https://goldenagri.com.sg/id/palm-oil-role-in-feeding-the-world/.

203. Mitropoulos, Angela, *Pandemonium: Proliferating Borders of Capital and the Pandemic Swerve*, VAGABONDS Series (London and New York: Pluto, 2020).

204. Dyett, Jordan and Cassidy Thomas, "Overpopulation Discourse: Patriarchy, Racism, and the Specter of Ecofascism," *Perspectives on Global Development and Technology* 18, no. 1–2 (2019): 205–24.

205. Patel, Raj, *Stuffed and Starved: The Hidden Battle for the World Food System* (updated edition) (New York: Melville House, 2012).

206. Zuckerman, *Planet Palm*, p. 162.

207. Schorb, Friedrich, "Fat as a Neoliberal Epidemic: Analyzing Fat Bodies through the Lens of Political Epidemiology," *Fat Studies* 11, no. 1 (2021): 1–13; For an informed free market critique of such taxes, see Snowdon, Christopher, "The Proof of The Pudding: Denmark's Fat Tax Fiasco" (London: Institute of Economic Affairs, 2013), https://iea.org.uk/sites/default/files/publications/files/The%20Proof%20of%20the%20Pudding.pdf.

208. Spratt, Tanisha Jemma Rose. "Understanding 'Fat Shaming' in a Neoliberal Era: Performativity, Healthism and the UK's 'Obesity Epidemic.'" *Feminist Theory*, 2021, 146470012110483.

209. Schorb, "Fat as a neoliberal epidemic."

210. See Strings, Sabrina, *Fearing the Black Body: The Racial Origins of Fat Phobia* (New York: New York University Press, 2019).

211. See de Angelis, Massimo, *The Beginning of History: Value Struggles and Global Capital* (London and New York: Pluto, 2006).

212. Chan, Kam Wing and Xiaxia Yang, "Internal Migration and Development: A Perspective from China," in *Routledge Handbook of Migration and Development*, ed., Tanja Bastia and Ronald Skeldon (London and New York: Routledge, 2020), pp. 567–84.

213. Wang, Ya Ping, Yanglin Wang, and Jiansheng Wu, "Housing Migrant Workers in Rapidly Urbanizing Regions: A Study of the Chinese Model in Shenzhen," *Housing Studies* 25, no. 1 (2010): 83–100.

214. Chan, Chris King-chi, "Community-Based Organizations for Migrant Workers' Rights: The Emergence of Labor NGOs in China," *Community Development Journal* 48, no. 1 (2012): 6–22.

215. See Solt, George, *The Untold History of Ramen: How Political Crisis in Japan Spawned a Global Food Craze* (Berkeley CA and London: University of California Press, 2014).

216. Cam, Lisa, "What's the Story behind Instant Ramen Noodles—and How Did Post-War America Influence Their Invention?" *South China Morning Post*, April 2020, https://scmp.com/magazines/style/news-trends/article/3077785/whats-story-behind-instant-ramen-noodles-and-how-did.

217. Solt, *The Untold History of Ramen*.

218. Gibson-Light, *Ramen Politics*.

219. Matory, *The Fetish Revisited*.

220. Gibson-Light, *Ramen Politics*.

221. Ibid., p. 204.

222. Ibid., pp. 200–201.

223. Gilmore, *Golden Gulag*.

224. Harvey, David, *Limits to Capital*, Essential David Harvey Series (London and New York: Verso, 2018).

225. Bhattacharyya, Gargi, *Rethinking Racial Capitalism: Questions of Reproduction and Survival* (London and New York: Rowman and Littelfield, 2018).

226. Robinson, William I., and Yousef K. Baker, "Savage Inequalities: Capitalist Crisis and Surplus Humanity," *International Critical Thought* 9, no. 3 (2019): 1–18.

227. Bernards, Nick, "'Latent' Surplus Populations and Colonial Histories of Drought, Groundnuts, and Finance in Senegal," *Geoforum*, 2019.

228. Patel and Moore, *The History of the World in Seven Cheap Things*.

229. Pye, "A Plantation Precariat."

230. Bhattacharyya, *Rethinking Racial Capitalism*.

231. Ibid.

232. Shilliam, Robbie, *Race and the Undeserving Poor: From Abolition to Brexit* (New York: Columbia University Press, 2018).

233. Azeri, Siyaves, "Surplus-Population and the Political Economy of Fear," *Critical Sociology* 45, no. 6 (2019): 889–905; Dyett and Thomas, "Overpopulation Discourse,"

234. Kundnani, Arun, "The Racial Constitution of Neoliberalism," *Race & Class* 63, no. 1 (2021).

235. Brown, Wendy, "Sacrificial Citizenship: Neoliberalism, Human Capital, and Austerity Politics," *Constellations* 23, no. 1 (2013). For fascinating exploration, see Wang, Keren, *Legal and Rhetorical Foundations of Economic Globalization: An Atlas of Ritual Sacrifice in Late-Capitalism* (London and New York: Routledge, 2021).

236. Agamben, Giorgio, *Homo Sacer: Sovereign Power and Bare Life*. trans. Daniel Heller-Roazen (Stanford CA: Stanford University Press, 1998).

237. Mbembe, Achille. "Necropolitics," *Public Culture* 15, no. 1 (2003): 11–40.

238. See Bremmer, *The Strange World of Human Sacrifice.*

239. Tatlock, Jason, "Human Sacrifice and Propaganda in Popular Media: More Than Morbid Curiosity," *Dialogue: The Interdisciplinary Journal of Popular Culture and Pedagogy* 6, no. 1 (2019).

240. Bremmer, Jan N, "Human Sacrifice: A Brief Introduction," in *The Strange World of Human Sacrifice*, ed., Jan. N Bremmer (Leuven: Peeters, 2007), pp. 1–10.

241. See Watts et al., "Ritual Human Sacrifice Promoted and Sustained the Evolution of Stratified Societies."

242. Sheils, Dean, "A Comparative Study of Human Sacrifice," *Cross-Cultural Research* 15, no. 4 (1980): 245–62.

243. See Watts et al., "Ritual Human Sacrifice Promoted and Sustained the Evolution of Stratified Societies."

244. For a fascinating exploration of how traditions of East African sacrifice today express themselves through a capitalist economic frame, see Atekyereza, Peter Rwagara, Justin Ayebare, and Paul Bukuluki, "The Economic Aspects of Human and Child Sacrifice," *International Letters of Social and Humanistic Sciences* 41 (2014): 53–65.

245. Hayek, F. A., The Road to Serfdom: Text and Documents, ed., Bruce Caldwell (Chicago and London: University of Chicago Press, 2007); Fukayama, Francis, *The End of History and the Last Man* (New York: Perennial, 1992).

246. See Martin, Randy, *Knowledge LTD: Towards a Social Logic of the Derivative* (Philadelphia: Temple University Press, 2015).

247. Gates, *The Road Ahead*; McGoey, *No Such Thing as a Free Gift*.

248. Docherty, *Blood and Bronze*, pp. 39–69.

249. Martin, Randy, *Empire of Indifference: American War and the Financial Logic of Risk Management* (Durham NC: Duke University Press, 2007).

250. Taleb, Nassim Nicholas, *The Black Swan: The Impact of the Highly Improbable*, second edition (New York: Random House, 2010).

251. LiPuma, Edward, and Benjamin Lee, *Financial Derivatives and the Globalization of Risk* (Durham, NC and London: Duke University Press, 2004).

252. Roy, Ananya, "Subjects of Risk: Technologies of Gender in the Making of Millennial Modernity," *Public Culture* 24, no. 1 66 (2012): 131–55.

253. Wu Ming 1, *La Q Di Qomplotto. QAnon e Dintorni. Come Le Fantasie Di Complotto Difendono Il Sistema* (Rome: Alegre, 2021).

254. Russonello, Giovanni, "QAnon Now as Popular in U.S. as Some Major Religions, Poll Suggests," *New York Times*, May 2021, https:// nytimes.com/2021/05/27/us/politics/ qanon-republicans-trump.html.

255. Federici, Silvia, *Witches, Witch-Hunting, and Women* (Brooklyn: Common Notions, 2018).

256. Wynter and McKittrick, "Unparalleled Catastrophe for Our Species?"

257. Gill and Taussig, *Becoming Palm*.

258. Pye, "Commodifying Sustainability."

259. Meijaard et al., "Oil Palm and Biodiversity."

260. Moore, Jason W., "Cheap Food & Bad Money: Food, Frontiers, and Financialization in the Rise and Demise of Neoliberalism," *Review*

*of the Fernand Braudel Center* 33, no. 2–3 (2010): 225–61.

261. Robins, *Oil Palm*, Chapter 8.

262. Ibid., pp. 348–50.

263. Pye, Oliver, Fitri Arianti, Rizal Assalam, Michaela Haug, and Janina Puder, "Just Transition in the Palm Oil Industry," *Transnational Palm Oil Labor Solidarity* (blog), September 2021, https://palmoillabor.network/just-transition-in-the-palm-oil-industry-a-preliminary-perspective/.

Thanks to our Patreon subscriber:

*Ciaran Kane*

Who has shown generosity and comradeship in support of our publishing.

Check out the other perks you get by subscribing to our Patreon – visit patreon.com/plutopress.

Subscriptions start from £3 a month.

**The Pluto Press Newsletter**

Hello friend of Pluto!

Want to stay on top of the best radical books
we publish?

Then sign up to be the first to hear about our
new books, as well as special events,
podcasts and videos.

You'll also get 50% off your first order with us
when you sign up.

Come and join us!

Go to bit.ly/PlutoNewsletter